Information Materials

Manuel Kretzer

Information Materials

Smart Materials for Adaptive Architecture

Springer

Manuel Kretzer
Chair for Computer Aided Architectural
Design (CAAD)
Institute of Technology in Architecture
Zurich
Switzerland

ISBN 978-3-319-35148-3 ISBN 978-3-319-35150-6 (eBook)
DOI 10.1007/978-3-319-35150-6

Library of Congress Control Number: 2016945987

Printed on acid-free paper

This Springer imprint is published by Springer Nature
The registered company is Springer International Publishing AG Switzerland

Foreword

Information Materials: Resonant, Potent Architecture

What kind of cities could result if we used the hybrid materials and new building systems described within *Information Materials*? The renewed 'softness' that Manuel Kretzer describes within this book evokes surfaces and structures of buildings within a city that follow densely layered forms of diffusive, deeply interwoven materials expanding and contracting, interacting with their surroundings. These building materials can be conceived as filters acting in oscillations: catching, harvesting, pulling, both expanding human influence and in turn expanding the impact of the world on us. Networks organized in loops and reflexes build up a deeply layered, deeply fissured set of relationships in which there are multiple sensitive boundaries. A new city designed to easily handle unstable conditions of cooling and heating in rapid cycles of shedding heat and warming and collecting heat again might look like a hybrid forest where each building is made from dense layers of ivy-like filters and multiple overlapping levels of porous openings. Like reticulated snowflakes or the microscopic manifolds of mitochondria, intense reticulation could form increased surface areas that make their reaction-surfaces potent.

Paths to industrial application and commercialization are launched by numerous examples of highly involved cycles of hand and machine-based manipulations, accompanied by cycles of virtual projection and physical prototyping. Key information materials are described including aerogels, bioluminescent materials, bioplastics and dye-sensitized solar cells, electroluminescent displays, electroactive polymers, soft robotics, and thermochromics. An appendix supporting application of the information with contact information for material suppliers is included.

In a section devoted to historical precedents and theory, Kretzer reveals the turbulence that has accompanied contemporary discussions of materiality. Marx's famous invocation '...all that is solid melts into air...' implies a revolution reacting to massive foundations and the framing of permanent construction. Two centuries of synthesis within the Industrial Revolution have left the world transformed, and yet that solid foundation still lingers. Kretzer offers nuanced readings of new paradigms positioned to offer distinctly new qualities. He includes Patrik Schumacher's optimistic conception of Parametricism alongside the early nineteenth-century Durand's systematic and encyclopedic treatment of architectural types, and in turn raises profoundly transformative visions that include Ludger Hovestadt and Vera Buhlmann's 'Metalithikum', the conception that solar energy can be considered an effectively infinite resource, bounded only by the limits of human intellect. In answer to the poignant political implications of such conceptions, Kretzer sets out a provocative model of effective public action framed by the social organization of the 'Materiability Research Network', a Web-based open-source participatory community designed for sharing information on fabrication and material synthesis. Ethical discussions accompany the practical application of Creative Commons licensing, offered by Kretzer as a legal framework that can support the evolving organization and collective contributions of an information commons.

A synthesis proposing a renewed conception of architecture integrated with material performance entitled 'Toward A New Softness' is positioned as a conclusion. The parallel and intertwined disciplines of applied craft, history, and analysis offer a compelling example of effective contemporary architectural practice. The work is clearly written, beautifully documented with unusually rich and precise visual documentation, and eminently readable. The clarity, focus, and special position of the author as an original creator and an activist within this emerging body of work positions this work as a practical manual dedicated to contemporary design craft, of interest both to professional and broad publics.

Philip Beesley

Preface

Very slowly, over thousands of years, the ideal of the perfect machine was approached—that ideal which had once been a dream, then a distant prospect, and at last reality: No machine may contain any moving parts.

Arthur C. Clarke, *The City and the Stars*, 1956, p. 161.

The content presented in this publication is an adapted version of my doctoral thesis that I completed at the Chair for Computer Aided Architectural Design, ETH Zurich, in early 2016. Working on my dissertation and on this book has been an exceptional, occasionally exhausting, but always an exciting journey. The knowledge that I gained on the field of *information materials* not only made me broaden my horizon and experience many previously undisclosed technologies, but especially introduced me to a myriad of fascinating individuals with whom I had the opportunity to exchange or collaborate, each of them mentioned in the *Credits* section at the end of this book.

My interest in this field began in 2009, when working at the office of Philip Beesley in Toronto, Canada. Employing shape-memory alloys for the actuation of his 'Breathing Pores,' made me wonder about the fascinating, very particular behavior of these dynamic materials, which perform in absolute silence not requiring anything but electricity. The thin wires, which are miraculously able to contract when heated, embedded in the untold beauty of Beesley's extraterrestrial installations, propelled me to question my experience in mechanical systems and convinced me that there had to be a softer, more organic side to mechanics and especially architecture.

After returning to Europe, from Canada, I had the privilege to receive a teaching and research position at Ludger Hovestadt's Chair for Computer Aided Architectural Design, ETH Zurich, in early 2010. Once again I was blown away, this time not by a certain materiality, but rather by the immense powers of information technology and their architectural application. The work that is conducted at the Chair in terms of programming, computation, and fabrication, and the people who do so, is outstanding and can hardly be compared to anything else I have seen since. Operating in such a vibrant environment had been a continuous inspiration and challenge and allowed me not only to discover and understand the possibilities of code and algorithms but also to freely develop my own agenda.

When commencing my dissertation in late 2011, my aim was to merge and unite these two experiences, the physical world of smart materials and the immaterial world of computers, which I refer to as *information materials*. A crucial part of this effort involved in extending my scope beyond the domain of architecture and making me delve into the areas of engineering, materials science, or even biology, disciplines of which I only had the faintest knowledge, but each provided a plethora of inspirations and opportunities for creative design.

Now, roughly 6 years later, I am more than convinced by the immense possibilities of employing *information materials* into architecture and design for the creation of adaptive buildings, spaces, products, and experiences. Yet evermore I am continuously fascinated by new ways of working, fostered through cross-disciplinary exchange and collaboration, driven by optimistic curiosity, and encouraged through models of openness and sharing. With such prospects, I am highly looking forward to continuing this work in the years to follow.

Zurich, Switzerland Manuel Kretzer
2016

Reference

Clarke AC (1956) The city and the stars. Orion Publishing Group, London

Contents

1 Current Global Challenges, the Concept of Adaptive Architecture, and the Possibilities of Information Materials . . . 1
1.1 Introduction . . . 1
1.2 Global Developments and Their Relevance for the Architectural Discourse . . . 3
1.2.1 Technological Advancements . . . 3
1.2.2 Demographic Growth . . . 4
1.2.3 Socio-anthropological Trends . . . 5
1.3 Challenges for Architecture and Building . . . 6
1.3.1 Junkspace or the Loss of Architectural Identity . . . 8
1.3.2 Superficiality and the Digitization of Materiality . . . 8
1.4 Flexibility as a Spatial Method to Address Change and Progress . . . 9
1.4.1 Adaptive Architecture . . . 11
1.4.2 Types of Adaptive Architecture . . . 12
1.4.3 Material Intelligence . . . 12
1.5 Information Materials . . . 13
1.5.1 Potential of Information Materials for Architectural Usage . . . 15
1.5.2 Shortcomings of Current Information Material Usage . . . 16
1.6 Toward a New Softness . . . 17
1.6.1 A Different Kind of Material Thinking . . . 18
1.6.2 Structure of the Book . . . 19
1.6.3 Line of Argumentation . . . 19
References . . . 21

2 The Ever-Changing Nature of Materiality and the Meaning of Materials in Architecture and Construction . . . 25
2.1 Introduction . . . 25
2.1.1 Structure of the Present Chapter . . . 27
2.2 Natural Materials . . . 29
2.2.1 The Influence of Vitruvian Thought on Current New Materialism Positions . . . 30

2.2.2 The Growing Detachment of Form from Materiality . . . 32
2.2.3 The Domination of Human Intellect Over Natural Materiality . . . 34
2.3 Industrial Materials . . . 36
2.3.1 Durand's Grammars in Relation to Schumacher's Parametricism . . . 38
2.3.2 The Intellectual Emancipation of Form from Materiality . . . 39
2.3.3 Technological and Material Advancement During the Industrial Revolution . . . 40
2.3.4 Functionality Takes Precedence Over Aesthetics and Form . . . 41
2.4 Synthetic Materials . . . 42
2.4.1 The Age of Plastics . . . 43
2.4.2 Utopian Visions for a Better Tomorrow . . . 44
2.5 Intermediate Summary and Conclusion . . . 45
2.6 Digital Materials . . . 46
2.6.1 The Emergence of Digital Design and Fabrication Techniques . . . 48
2.6.2 New Instruments for the Architectural Design Practice . . . 49
2.6.3 The Digital Chain: From Design to Production . . . 51
2.7 Information Materials . . . 52
2.7.1 Programmable Matter . . . 52
2.7.2 The Soft Kinetics of Smart Materials . . . 53
2.7.3 Definitions of Smart Materials . . . 54
2.7.4 Concepts for Information Materials Usage in Architecture . . . 56
2.7.5 Material Behavior in Regards to Energy, Time, and Space . . . 58
2.8 Final Summary and Conclusion . . . 60
References . . . 62

3 Materiability—An Attempt for the Education of an Information Material Literacy in Respect to Emerging Materials . . . 67
3.1 Introduction . . . 67
3.2 The Process of Physical Making . . . 70
3.2.1 The Inherent Dynamics of Information Materials . . . 71
3.2.2 The Educational Framework of the Materiability Approach . . . 71
3.3 Access to Information . . . 73
3.3.1 Approaches for the Mediation of Material Knowledge . . . 75
3.3.2 The Materiability Research Network: Access to Information Materials . . . 76

3.4 Information Material Literacy 77
3.4.1 Collective Making in the Third Industrial Revolution 78
3.4.2 The Materiability Research Network: A Growing Community 79
3.5 Summary and Conclusion 81
References 82

4 A Selection of Emerging Information Materials, Their Properties, Fabrication, and Application in Speculative Spatial Installations 85
4.1 Introduction 85
4.2 Aerogels 86
4.2.1 History 86
4.2.2 Material Structure and Synthesis 87
4.2.3 Fabrication 90
4.2.4 Applications 95
4.3 Bioluminescence 95
4.3.1 History 96
4.3.2 Material Structure and Operation 96
4.3.3 Fabrication 99
4.3.4 Applications 101
4.4 Bioplastics 102
4.4.1 History 102
4.4.2 Material Structure and Synthesis 103
4.4.3 Fabrication 106
4.4.4 Case Studies 108
4.4.5 Applications 110
4.5 Dye-Sensitized Solar Cells 111
4.5.1 History 112
4.5.2 Material Structure and Operation 112
4.5.3 Fabrication 113
4.5.4 Case Studies 116
4.5.5 Applications 119
4.6 Electroluminescent Displays 119
4.6.1 History 120
4.6.2 Material Structure and Operation 121
4.6.3 Fabrication 122
4.6.4 Case Studies 126
4.6.5 Applications 129
4.7 Electroactive Polymers 130
4.7.1 History 130
4.7.2 Material Structure and Operation 131
4.7.3 Fabrication 132
4.7.4 Case Studies 135
4.7.5 Applications 137
4.8 Soft Robotics 138
4.8.1 Material Structure and Operation 139
4.8.2 Fabrication 141

4.8.3 Case Studies 143
4.8.4 Applications 144
4.9 Thermochromics 145
4.9.1 History 145
4.9.2 Material Structure and Operation 146
4.9.3 Fabrication 148
4.9.4 Case Studies 152
4.9.5 Applications 159
4.10 Summary and Conclusion 160
References 161

5 Towards a New Softness 167
5.1 Introduction and Book Summary 167
5.2 Thinking Information Materials 169
5.2.1 How Has Material Development and Usage Evolved in Architecture Over the Course of Time and How Has This Influenced the Practice and Education of the Architect? 169
5.2.2 What Is the Current Meaning and Usage of Information Materials in Architecture and How Might It Change in the Future? 170
5.2.3 What Are Didactic Models that Focus on the Mediation of Dynamics, Processes, and Behaviors and How Can Explorative Material Research Become a Vital Aspect of Architectural Education? 171
5.2.4 How Can a Distinct Pedagogic Approach Be Evaluated and How Can Research Results Be Disseminated to a Large, Global Audience? 172
5.3 Liquid Architecture 173
5.4 Towards a New Softness 175
References 178

Appendix A 181

Glossary 189

Current Global Challenges, the Concept of Adaptive Architecture, and the Possibilities of Information Materials

1

1.1 Introduction

> We are the slaves of our technical improvement and we can no more return a New Hampshire farm to the self-contained state in which it was maintained in 1800 than we can, by taking thought, add a cubit to our stature or, what is more to the point, diminish it. We have modified our environment so radically that we must now modify ourselves in order to exist in this new environment. We can no longer live in the old one. Progress imposes not only new possibilities for the future but new restrictions.
>
> Wiener (1954, The Human Use of Human Beings, p. 46)

When following the current architectural debate one can observe a continuously rising demand for new design and construction strategies which are largely aiming at improving a building's energy efficiency during its erection and life-cycle. The necessity for such change is affirmed by apparently unprecedented global events, such as ecological degradation, vanishing resources, or other environment-related issues. Considering the undeniable impact of architecture on nature, increased ecological sensibility is extremely important yet the polemics of how these transformations are often demanded should be evaluated with care, since they first are not new and second highly ambivalent.

One prominent example is the position of the mathematician and founding father of **cybernetic theory** Norbert Wiener, who more than 60 years ago explored how vast technological changes that according to him were "unexampled in earlier history" (Wiener 1954, p. 46), have and will continue to influence society, economy, and the environment. Wiener emphasizes that the effects humans have on nature are irreversible and that progress not only creates new opportunities but likewise limitations. In a fairly pessimistic tone he urges mankind to consider technology and its impact on nature in a more holistic and foresighted way and learn from and adapt to its possible implications. Advances in automation and machine intelligence, in which he sees great potential to facilitate and reduce human labor, will thus only be of lasting value if applied consciously.

A few years later, in 1962, published the marine biologist Rachel Carson *Silent Spring* raising questions about the impact of synthetic pesticides and fertilizers on natural ecosystems. Picturing a world devoid of animal and plant life she argues that chemicals have equally contaminating effects as nuclear bombs and thus might well lead to the long-term extinction of mankind. To prevent such damage she demands increased control in spreading man-made compounds in nature and above all a greater public awareness of their usage and consequences. Especially since their effects often go beyond influencing the target pests but "accumulate in the tissues of plants and animals and even penetrate the germ cells to shatter or alter the very material of heredity upon which the shape of the future depends" (Carson 1964, p. 13).

M. Kretzer, *Information Materials*, DOI 10.1007/978-3-319-35150-6_1

Equally influential as Carson's work, which inspired much of the environmental movement in the industrial world, was Alvin Toffler's *Future Shock*. The book was published in 1970 and 2 years later released as a documentary film, narrated by Orson Wells. The visionary Toffler uses the term 'future shock' to describe a certain mental state of individuals or societies when being exposed to "too much change in too short of a time" (Toffler 1970, p. 1). The origins of such conditions, which he believes lead to stress and disorientation, are mainly to be found in increased social, demographic, and especially technological changes. Hence to prevent the extensive spread of future shock but also other related environmental and ecological tragedies he demands "the conscious regulation of technological advance" (Toffler 1970, p. 428), which, in junction with further political and educational control strategies, will restore order and stability.

Interestingly the basis and line of argumentation of these three, and myriad other speculations about the future, are fairly similar, departing from certain global, mostly environmental tendencies, which, if they continue to progress as they have in the past, will culminate in some sort of disastrous, potentially apocalyptic climax. In order to avoid or at least lower their cause, humanity thus either has to radically change its collective behavior and/or accept the introduction of other far-reaching strategies. Yet just as diverse, subtle, or totalitarian such confrontations are, the exact same story can obviously also be used to conclude in the opposite, highly optimistic visions. During the hype of the Industrial Revolution for example, the French artist Villemard published a set of postcards on how he imagined Paris in the year 2000 to be like (Figs. 1.1 and 1.2), including new types of urban life and transportation.

Similarly colorful were utopian ideas during the 1950s and '1960s. While Wiener and Toffler demanded a reconsideration of technological progress, the artists James R. Powers or Klaus Bürgle (Fig. 1.3) envisioned the absolute inverse, a materialistic consumer culture thriving in a lucent environment comprised of organically shaped buildings, vibrant urban spaces, and new types of public and personal transportation systems (Kretzer 2014a, b, p. 502).

Fig. 1.1 The French artist Villemard had already at the beginning of the twentieth century fairly interesting visions on urban life and new kinds of aerial transportation (En L'an 2000, Une Station de Taxis Volants, Villemard 1910)

Fig. 1.2 Considering current trends in robotic construction, some of Villemard's visions weren't too far-fetched. (En L'an 2000, Chantier de Construction Électrique, Villemard 1910)

Fig. 1.3 The German artist Klaus Bürgle imagined new types of public transportation systems within thriving urban spaces (Der Verkehr der Zukunft aus das neue Universum '76, Michael Peters for Klaus Bürgle 1959, www.retro-futurismus.de)

Whereas today's images of the future might be a little less expressive, in both optimistic and pessimistic manners, the challenges, and

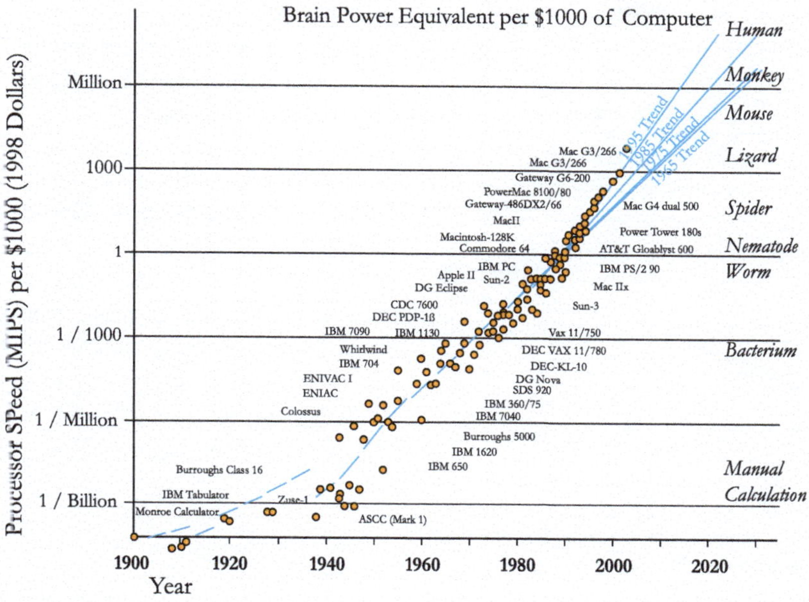

Fig. 1.4 Evolution of computer power in relation to cost, 1900–1998 (adapted from Hans Moravec)

opportunities, especially for architecture, are just as manifold as they always have been. To provide a brief overview which can be used as a basis for further discussion the following sections present and evaluate three persisting tendencies for their architectural relevance, *Technological Advancements, Demographic Growth,* and *Socio-anthropological Trends.*

1.2 Global Developments and Their Relevance for the Architectural Discourse

1.2.1 Technological Advancements

The American author, futurist, and director of engineering at Google, Ray Kurzweil, expands upon Moore's Law, which describes the bi-annual doubling of transistors on integrated circuits, and hence is used to predict the further evolution of computing hardware. From this, he develops his *law of accelerating returns* which prognoses that the rate of change in various systems, especially technological ones, increases exponentially. Kurzweil graphically reinforces his theory by analyzing the growth rate of numerous significant developments during the twentieth and early twenty-first century, such as the increase in computer processor performance, the rising amount of Internet hosts, as well as non-technological phenomena like the growing investment in education and learning. Relating to a chart by Hans Moravec, which compares the evolution of computational performance to organic intelligence (such as bacteria, worms, spiders, lizards, etc.), Kurzweil forecasts that by 2020 computing power will be equivalent to human brain capability as depicted in Fig. 1.4 (Kurzweil 2005, pp. 56–57).

Building upon this assumption he argues further that around 2045 this will lead to, what he calls, a technological singularity (Fig. 1.5) "representing a profound and disruptive transformation in human capability" (Kurzweil 2005, p. 136). Once this point has been surpassed mankind will have the capacity to merge with intelligent machines and through this surmount the biological restraints of its physical and psychological constitution.

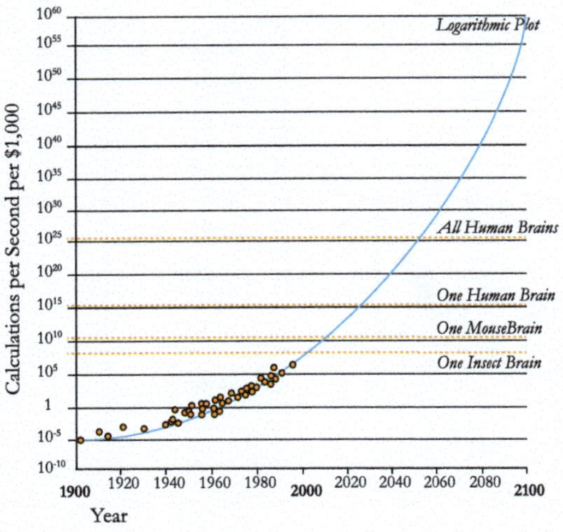

Fig. 1.5 Exponential growth of computing twentieth through twenty-first century (adapted from Ray Kurzweil and Kurzweil Technologies Inc. 2005)

Humans will then not only be able to severely, maybe infinitely, extend their lives but also be able to enhance their minds by genetic alterations, nanotechnology, and artificial intelligence. This will eventually lead to the creation of new (hybrid) life forms that will be far superior to all of today's human intellect combined (Kurzweil 2005, p. 9).

Kurzweil's vision, due also to its scientific and statistically grounded format, has meanwhile established a vastly growing international following from various backgrounds. The core of his prognosis, the accelerating development of new technologies, in terms of their amount, ubiquity, and power, likewise raises a number of interesting questions for architecture and building, such as:

- Can systems that posses a certain intelligence or autonomy be embedded into buildings without suggesting a loss of control and safety?
- If humans become digitally enhanced hybrids, what kinds of services will architecture have to provide other than shelter and privacy and how will these be materialized?
- How can continuously changing technologies be incorporated in a building's technical infrastructure without requiring fundamental updates every few years or even months?
- What are methods to educate and provide access to such developments in a productive manner while avoiding falling into retrogressive dogmatics?

1.2.2 Demographic Growth

In the past 50 years the world population has more than doubled, growing from 3 billion to about 7.3 billion today and, according to a 2013 UN report, will be reaching an estimated peak of 9.6 billion by the end of 2050, as displayed in Fig. 1.6 (UN New Centre 2013).

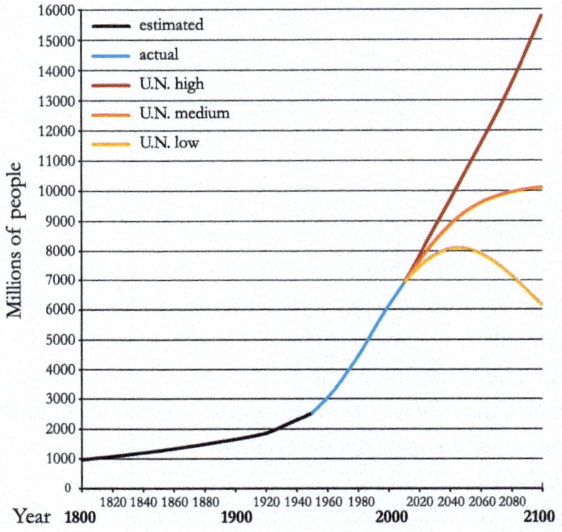

Fig. 1.6 World population chart from 1800 to 2100, showing both estimates and actual population counts (adapted from Cobb 2008)

This dramatic growth in population will be accompanied by further demographic changes, such as an increase in average life expectancy, which has already risen from 52 years of age in 1962 to 71 in 2013 and will probably top 76 in 2050 (Central Intelligence Agency). In addition more and more people will continue to be drawn to urban areas, meaning that by mid century over 70 % of the planet's population is expected to have moved to vastly growing metropolises (CBSNEWS 2008) (Fig. 1.7).

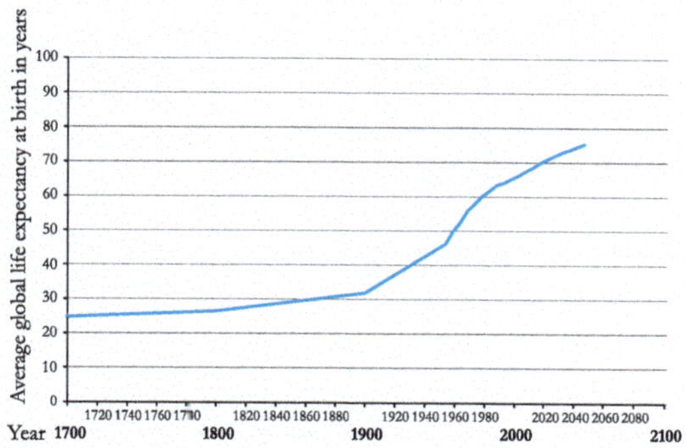

Fig. 1.7 Average global life expectancy at birth from 1700 to 2100 (Kretzer 2015)

As humans who live and work in urban places perform over 90 % of their activities within enclosed spaces (Klepeis et al. 2001, p. 231) and spend about 15 h per day at home (Brasche and Bischof 2005, p. 248), this will not only lead to a significant increase and densification of built volume but also a gradual deterritorialization of rural and less developed regions.

Further predictions assume that by 2050 humanity will need 80 % more energy than today (OECD 2012), and, due to an increasingly sedentary lifestyle, by 2030 two-thirds of the global population will be obese (Airbus). At the same time, however birth rates in many developed countries are decreasing, which especially within Europe will lead to a decline in population posing utterly different challenges than for example on the African continent, whose people are expected to more than double within the next 40 years to about 2.4 billion, representing a quarter of the then world population. For architecture and urban planning the challenges are pressing and far more complex than the following questions. Yet especially in relation to energy and materials a number of interesting topics should be addressed:

- With cities being responsible for some 75 % of the global energy consumption and 80 % of greenhouse gas emissions how can air and water quality be improved while reducing energy use and waste production (United Nations 2007)?
- What are clean alternatives for energy production and how can these become integral and not just additional parts of architecture?
- How can space be used rationally without decreasing its comfort, quality, and affordability?
- What materials and techniques would allow the construction of lighter, more efficient, and maybe even transformable or responsive structures?

1.2.3 Socio-anthropological Trends

Children born after 2010 are unofficially referred to as Generation Alpha. Since to date not much data exists on the demands of these young people, assumptions on their future are usually based on studies of Generation Y, who are born in between 1980 and 1994 and Generation Z, who are born from 1995 to 2010 (Kretzer 2011, p. 165).

Generation Y grew up in a media-saturated, brand conscious environment, very different to previous times. Their childhood was strongly influenced by the birth of digital technologies, such as the Internet, Email, Cable TV, the Personal Computer, and the Laptop. They are quickly adapting to new developments and believe they make life easier, bring family and friends closer together, and allow for more efficient time management (Kohut et al. 2010, p. 26).

Generation Z was born into smaller families with the fewest siblings of any era. While parts of Generation Y still have a strong recollection of times before mobile phones and the World Wide Web, Generation Z are true 'Digital Natives.' As part of a global study among teen online community and social networking service 'Habbo Hotel,' 55 % of those polled believed that digital content will soon replace traditional print media, 34 % supported the idea of expanding our linguistic usage with the informal language developed in chat and text messaging, and over a third felt that physical meetings could be superseded by online interactions within virtual homes (Music Industry News Network).

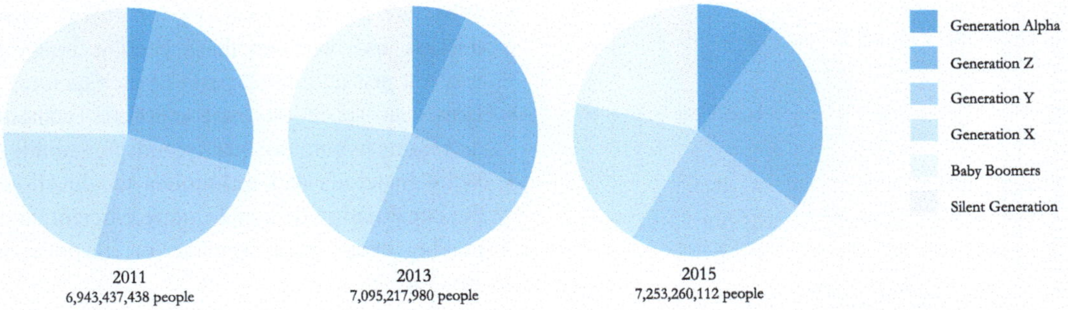

Fig. 1.8 Demographic distribution of percentage of world population by generation in the years 2011, 2013, and 2015 (Kretzer 2015)

Generation Alpha currently makes up about 13 % of the world population and according to Australian demographer Peter McDonald might become the biggest generation yet (ABC News). Their growth over the past 5 years in relation to other generations is graphically displayed in Fig. 1.8. 'Alphas' are expected to go to school earlier, study longer, work more, follow a multitude of different careers, and change their employer more often. Being the children of older and wealthier parents they are anticipated to become more materialistic and technology-focused than any of their preceding generations, owning less physical objects at a time, which they will however change more frequently (McCrindle 2008, p. 1). Due to constant connectivity they will exhibit an immanent urge to stay up to date while expressively demarcating their identity within a largely digital culture (Palfrey and Gasser 2008, pp. 125–126).

For the creation of future spaces, both in private and work-related environments, a number of questions arise:

- What are design concepts that build upon the demand of current and future generations for independence, self-expression, and the celebration of their individuality and diversity?
- How can areas be created that positively address the amalgamation of virtual and physical worlds while retaining the necessary stability and security architecture stands for?
- What technologies, systems, and materials have potential for the establishment of such multidimensional spaces and how can they become integral architectural elements?
- How can new job and educational models that are most likely much more flexible and ephemeral than existing structures, be translated into productive environments that foster a healthy life-work balance?

1.3 Challenges for Architecture and Building

Of course such predictions and conclusions, no matter how statistically grounded they might appear, can only provide hints and indications of possible future scenarios. Yet it remains important to take them seriously and consider their relevance for the present and future architectural discourse, not only for their practical value but also in terms of farther-reaching, more speculative implications (Kretzer 2014a, b, p. 464). A simple example to visualize such thinking is photovoltaic technology. As Fig. 1.9 displays the average cost of solar panels has exponentially decreased from 76.67 USD/W in 1977 to 0.36 USD/W in 2014. At the same time the amount of worldwide solar photovoltaic capacity has increased to almost 140.000 MW (Fig. 1.10), which is already enough to power every home in Germany (Earth Policy Institute 2014).

While such a tendency might at first sight simply convince people of the potentials of green energy and the reduction of CO_2 emission it can

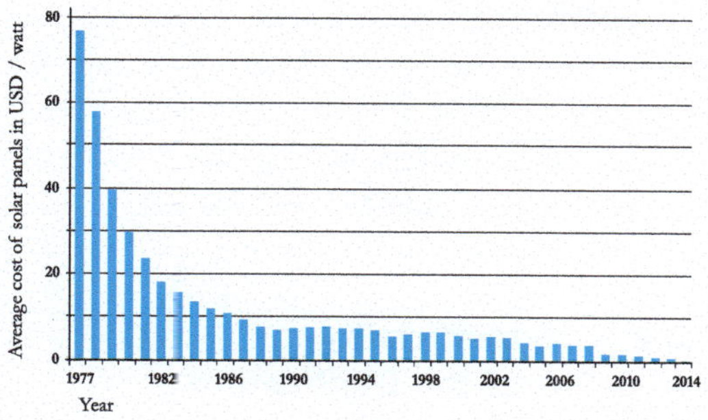

Fig. 1.9 Price history chart of crystalline silicon solar cells in USD/W since 1977 (adapted from Fassbind 2015)

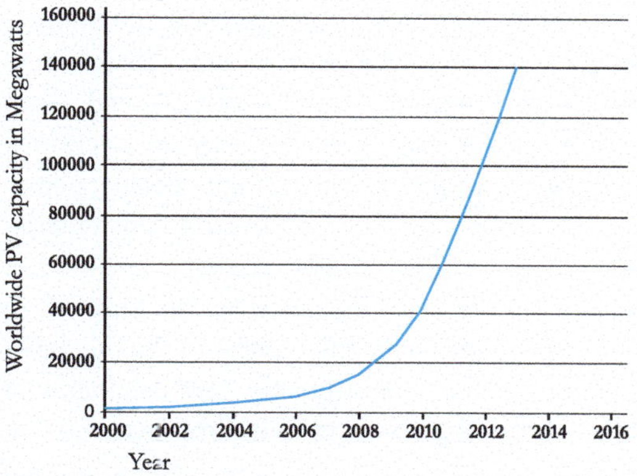

Fig. 1.10 World cumulative solar photovoltaics installations 2000–2013 (adapted from Matt Roney, Earth Policy Institute: China leads world to solar power record in 2013)

also be thought much more radically, as Ludger Hovestadt suggests.

The amount of solar radiation that reaches the planet every year is roughly 10,000 times higher than the annual global energy consumption. If mankind were thus able to only capture 0.01 % of this energy it would already have enough to cover any kind of power requirement today. Considering that both production costs of solar technology will continue to decrease while photovoltaic capacity will further grow, soon not only the total worldwide energy demand could be covered but actually much more than ever needed in the future. Suddenly thinking in terms of limited resources, which we are trained to by Wiener, Toffler, and the likes, turns into thinking in regard to abundant potentials, especially since solar cells are artificially created devices. The production of power as the base for human existence thus becomes no longer dependent on the dwindling supplies of our planet but turns into a matter of pure intellect.

In that sense the use and integration of technology into architecture changes into an intellectual question, which cannot be solved through classical materialistic thinking but instead requires a different approach, emphasizing its potentials and abilities (Hovestadt 2013, p. 65).

Such awareness becomes even more obvious when considering the ubiquity of technology and how seamlessly it is already integrated into our physical environment. Smartphones, wearable computers, augmented reality, or the Internet of Things are well-established terms that have and will continue to radically influence the way people move, interact, communicate, and behave. Thus, they have a significant impact on society, culture, and nature. Since architecture can, in a very basic sense, also be seen as an extension of the body, which provides first of all shelter, it is surprising that these trends have had so little effect on both spatial design and its structural manifestations, whereas other industries, like the aircraft, automotive, or entertainment area, are much more progressive and are shaped by and merge with the technologies they encompass. With cars for example, even though the basic principles have remained the same, their form, performance, and materiality have in the last 130 years evolved profoundly and so it seems that innovation is far more integrated and influential in both their conception and development (Kretzer 2015, p. 48).

The same can be said for telephones, computers, photo cameras, television screens, and any other consumer electronics devices. In architecture, however, albeit it is readily dependent on numerous technologies, they "have never really been incorporated in either the ideology or the theory of architecture" as Rem Koolhaas points out during an introduction to the 2014 Venice Biennale (Dezeen 2014).

1.3.1 Junkspace or the Loss of Architectural Identity

> If space-junk is the human debris that litters the universe, Junk-Space is the residue mankind leaves on the planet.
>
> Koolhaas (2002, Junkspace, p. 175)

Architecture's reluctance in absorbing new technologies can surely be understood, although not justified, from an economical perspective, what remains worrying however is that such unwillingness can in many cases already be felt during the early stages of conceiving design. Of course in all industries costs and expenses eventually determine which technology or material should be applied, yet this does not hinder but if anything propels speculation and experimentation. Koolhaas describes architecture's incompetence in responding to new developments and using them to create spatial meaning and identity by calling it 'Junkspace.' Junkspace portrays a generalized type of building, which according to him is appearing all over the planet, uniting everything into "a seamless patchwork of the permanently disjointed." Referring to Modernism as a rationale program driven by science and technological progress, Junkspace is the entropic offspring of disoriented global activity, the fatal triumph of quantity over quality.

Koolhaas' critique is that of a loss of value and meaning, of too many quick decisions without considering their larger implications, and of not appropriately addressing the very essence of architecture, the creation of space: "Architects could never explain space; Junkspace is our punishment for their mystifications." To Koolhaas architecture has become superficial and shallow, focusing on its formal appearance rather than its larger purpose. "Junkspace is sealed, held together not by structure but by skin, like a bubble" (Koolhaas 2002, p. 176).

Surely Koolhaas' opinion might just be as generalizing as the generalization he criticizes and by being the one naming it he elevates himself not only above all that is built but also beyond any type of counter-critique. Yet despite its polemic brevity and subjective perspective, he makes a number of points which can be seen in a more general, broader sense, turning superficiality, and standardization into a concept that goes beyond sole architectural associations.

1.3.2 Superficiality and the Digitization of Materiality

Considering the quickly changing nature of our environment and the vastly growing amount of everything one might argue that a certain superficiality and the introduction and adherence to norms and rules that allow filtering and comparing are the only way to keep an overview. Evaluating things based on their visual appearance or other easily comprehensible characteristics proves to be a fast and productive method of processing diverse information. In respect to architecture this can be witnessed in the large amount of databases, libraries, and catalogs that allow browsing through building materials in relation to specific quantifiable properties and applications. However, what tends to be forgotten or neglected is that any kind of standardization and categorization inevitably leads to loss of valuable information that at a later point can hardly be retrieved.

To Koolhaas this adds just another point to his argumentation. While to him materiality in the past represented a final state that could only be changed through partial demolition, materials today "have become more and more modular, unitary and standardized." The ever-decreasing tendency of turning materials into predigitized 'crypto-pixel,' only to, at a later point, reassemble them into usable units, is according to Koolhaas rooted in the cultural decision to abandon repetition and regularity and instead celebrate irregularity and uniqueness, again highlighting a preoccupation with form and appearance.

In an essay considering the reciprocal influence of form and material, architect and author Simone Jeska implies that the focus on surface rather than a material's intrinsic properties is directly linked to the introduction of CAD tools. In junction with new production methods these have resulted in architectural shape that is "robbed of its material dimension [and] only during manufacture [...] finds a reference back to the material." She further observes that the benefits of computer-controlled design and fabrication are reflected in new architectural terminologies, such as "transarchitecture, genetic architecture or flowing architecture." Yet, despite the dynamic notions of such expressions, they only remain flexible as long as they are kept within the virtual realm, which leads to the assumption that the concept of physical form has been replaced by the term digital design (Jeska 2008, p. 28). Professor at the Faculty of Architecture, TU Delft, Kas Oosterhuis, however insists that dynamic processes should under no means end with the completion of a building. He proposes the creation of so-called 'Hyperbodies,' which are "programmable building bodies that change their shape and content in real time." These will lead to an architecture which "will no longer remain static" but finally become liquid "not only as a metaphor in the design process, but in real life and in real time." Such buildings that are flexible and dynamic in their formal and structural constitution, would be much better equipped to address external as well as internal changes and challenges (Osterhuis 2002, pp. 42, 76).

1.4 Flexibility as a Spatial Method to Address Change and Progress

> Human beings are flexible creatures. We move about at will, manipulate objects and operate in a wide range of environments. [...] Most cultures now live a more or less sedentary life, but it could be that flexibility is once again becoming a priority in human development and that technological and economic changes are forcing, or at least encouraging a new form of nomadic existence.
>
> Kronenburg (2007, Flexible: Architecture that Responds to Change, p. 10)

Professor of Architecture, Robert Kronenburg from the University of Liverpool, highlights the benefits of flexible buildings, which he describes as "intended to respond to changing situations in their use, operation or location," in respect to contemporary cultural and social shifts. According to him, flexible buildings have various advantages over static ones, like being able to respond to a user's experience and input, possessing the power to adapt to technological innovation quicker, and being more economically and ecologically sound. On the example of Toyo Ito's Matsumoto Performing Arts Centre in Japan he explains four main strategies that mark the project's spatial flexibility. First, the building incorporates a number of transformable elements, such as a ceiling that can be lowered for different acoustic scenarios, a transformable stage, and adaptable seating to create different views and perspectives. Second, it features a large adaptable space, a fluid volume, which does not impose a particular function but rather stimulates a variety of uses and occupations. Third, the building encourages interactive operation across various levels, inviting visitors to exchange, and interact with each other. And finally, it includes numerous movable parts, mobile components, such as ticket desks, catering and concession stands, wardrobes, chairs and tables, etc. to quickly reorganize the space for varying purposes.

Flexible schemes, as the one described above, can help to alleviate problems through response and adjustment, yet the majority of today's buildings remain static entities derived from mass-produced elements and "built to the lowest common denominator for (apparently) standardized people, carrying out standard functions," as Kronenburg (2007, pp. 12, 85–87) observes. Yet obviously all people are distinct, having individual properties and requirements that hence, at least to some extend, need to be suppressed or adjusted to comply with existing systems. Clearly, such awareness is not new but has continuously occupied architectural thinking and practice throughout the centuries. A particular frustration with the status quo was boiling in the early 1960s, when professional decision-making seemed overwhelmed with the rising complexities of the built environment. While Robert Venturi fought the aesthetics of functionalism by demanding the return to classicist ornamentation and by focusing on the world of the consumer society (Kruft 1994, p. 440), Yona Friedman proposed the creative integration of the individual user into the design process. Friedman dissociates himself from anthropomorphic uniformity and "the concept of the 'average man' being a quantity of ideas and objects." Instead he suggests the programmatic development of an architectural framework that allows for flexibility, diversity, and change according to the inhabitant as a unique individual, each one different from the others, whose future behavior does not allow prevision.

In order to overcome traditional means of spatial expression and to democratize design incentives he promotes the development of a universal language in the form of cooking instructions, recipes, or tutorials that enable the inhabitant to experiment and evolve within a manageable framework (Friedman 2006, pp. 15, 37, 57). Such a system requires a user, who is willing to engage into the discourse and become physically active, a *creative* user, who according to the architectural historian Hill (2003, p. 27) is distinct from the *passive* or the *reactive* user.

> The passive user is predictable and unable to transform use, space and meaning. The reactive user modifies the physical characteristics of a space […] but must select from a narrow and predictable range of configurations largely defined by the architect. The creative user either creates a new space or gives an existing one new meanings and uses. Creative use can either be a reaction to habit, result from the knowledge learned through habit, or be based on habit, as a conscious, evolving deviation from established behavior.

Distancing himself from functionalist architecture, which emphasizes spatial determinism and the necessity of passive tenants, Hill refers to Adrian Forty, Professor of Architectural History at The Bartlett, London, when highlighting spatial flexibility as a concept for buildings to adapt

Fig. 1.11 Ant Farm Clean Air Pod at U.C. Berkeley Campus, Earth Day 1970. *Pictured* Andy Shapiro, Kelly Gloger (all rights reserved—Ant Farm 1970)

to changes and engage with their users on various levels. Forty classifies three strategies for architectural flexibility: by *technical methods*, by *spatial redundancy*, and as a *political strategy*.

The latter, *flexibility as a political strategy*, which is also the direction Friedman pursues, emerged as a criticism toward capitalism in the early 1960s, strongly promoted by Constant Nieuwenhuys and the Situationist International, a group who imagined modern society as an accumulation of deliberately spectacular and playful moments in time. The radical spirit of the collective inspired other artistic formations like Archigram, Haus Rucker Co., Coop Himmelb(l)au, or AntFarm (Fig. 1.11) who not only developed large Utopian visions of modern cities but also staged open happenings, actively engaging the public, to question society and propose anarchistic, anti-consumerist alternatives.

The second possibility, *flexibility by spatial redundancy*, assumes that a room is so big that it can easily contain a variety of uses, like for example the spaces in baroque palaces, which were not predetermined to certain actions. Hill goes on to describe the (modernist) open plan as yet another type of spatial redundancy, which distinct from *flexibility by technical means*, is less tied to physical adaptation of the space than to a change in perception of the consumer. As prominent precursors he mentions the traditional Japanese House, which inspired projects like the Barcelona Pavilion by Mies van der Rohe

Fig. 1.12 Mies van der Rohe's Barcelona Pavilion at the 1929 International Expo, best known for its large flat roof uniting the interior and exterior space (Kretzer 2016)

(Fig. 1.12) or Le Corbusier's Villa Savoye in Poissy.

Flexibility by technical means highlights the use of mechanical kinetic systems to transform space either in an automated or human-induced way. Early examples include the Rietveld Schröder House in Utrecht as well as the Maison de Verre in Paris, which both have various movable parts incorporated that allow the space to be divided and reconfigured. Referring to Forty, Hill identifies Cedric Price's Inter-Action Centre in London as "the closest built example of *flexibility by technical means* through demountable elements." He goes on and points out new technologies and their influence on architectural concepts, such as the rise of biotechnologies or cybernetics. Again looking at the work of Cedric Price he mentions the 'Generator Project,' a visionary, unfortunately unrealized proposal with cybernetician John Frazer for a retreat and activity center in Florida, as a key development "towards a building with intelligence." Such an adaptive system would be able to autonomously engage in the transformative process and thus become much more than just a mechanically reconfigurable space (Hill 2003, pp. 30–34).

1.4.1 Adaptive Architecture

> Architecture is considered as a form of artificial life, subject, like the natural world, to principles of morphogenesis, genetic coding, replication and selection. The aim of an evolutionary architecture is to achieve in the built environment the symbiotic behavior and metabolic balance that are characteristic of the natural environment.
>
> Frazer (1995, An Evolutionary Architecture, p. 9)

John Frazer, a pioneer in applying computer technologies in architecture, argues that architecture has become an essential part of the natural ecosystem and should therefore draw direct analogies from natural processes. Frazer proposes to achieve in the built environment a symbiotic behavior akin to characteristics found in nature. He emphasizes that this approach should reach beyond formal similarities but rather focus on nature's organic operating principles. Frazer goes on to highlight the benefits such a system would have for the environment, especially in regards to material usage, waste, energy consumption, and associated costs, but most of all for the user. In that respect he points to the unique ability of the aforementioned Generator Project, to "register its own boredom and make suggestions for its own reorganization." The Generator was intended to consist of 150 mobile cubes that could be rearranged by a crane to enhance the visitors' actions. If people did not request enough changes, or if the parts were not used properly, the system would begin proposing new configurations by itself, leading to unexpected and surprising constellations. This capacity, to learn from its own propositions, and not only suggest alternatives but also reevaluate them and evolve accordingly would essentially enable the building to develop a certain self-awareness, "of being able to anticipate the implications of its actions, as any good environmental control system should be able to do" (Frazer 1995, pp. 16, 41).

Michael Fox and Miles Kemp, authors of *Interactive Architecture*, stress the importance of John Frazer's work and criticize that evolutionary systems, and in particular genetic algorithms, are today still mainly used as formal design tools and not for the creation of adaptive and intelligent spaces (Fox and Kemp 2009, p. 243). Since numerous terminologies exist in order to describe buildings or spaces, which can dynamically adapt toward user presence, input, or behavior it is helpful to elaborate certain descriptions in more detail.

1.4.2 Types of Adaptive Architecture

A fairly clear definition of *interactive* architecture is provided by Tomasz Jaskiewicz, Assistant Professor in Industrial Design at TU Delft, describing it as "architecture that exhibits autonomous behavior, in which that behavior evolves through interactions with its users and environment" (Jaskiewicz 2013, p. 3). Interactive thus stands for a bilateral relationship between spaces or buildings and their inhabitants and a resulting change on both sides. London-based architect and artist Haque (2007, p. 26) relates to Gordon Pask's Conversation Theory, which is often mentioned as one of the earliest approaches toward interactive environments, when clarifying:

> Originally, interaction was distinguished from circular 'mutual reaction': it was about affecting not just actual output (in response to input) but also about affecting the way that output is calculated.

This demarcates *interactive* from *active*, which assumes a system that simply replies or reacts to user needs or desires. Jaskiewicz differentiates further between the terms *automated* architecture, which focuses on building automation systems, *smart* architecture, as a continuation of the automated building but with a certain artificial intelligence, *reactive* architecture, as a system with a preprogrammed behavior, *responsive* architecture, which wants to establish a dialogue between user and building and *transactive* architecture, which highlights the importance of the emergence of a transaction during such a dialogue. He continuous to distinguish the terms *performative* architecture, *kinetic* architecture, or *dynamic* architecture, which focus less on an exchange between the built and the user but rather articulate active structures in order to create particular effects (Jaskiewicz 2013, p. 43).

Notwithstanding the importance of clear definitions, Philip Beesley points out that "architecture has always been inventive and adaptable," independent of time, location, or technology. He also asserts that the current era offers unique potential in both technological advancements and societal and environmental challenges. In that sense, Beesley encourages looking beyond superficial effects and aesthetic sensationalism but rather thinking about the emotional impacts such an "expansion of the power of architecture" would have on its inhabitants (Beesley et al. 2013, pp. 1–2). This becomes especially interesting when considering adaptive systems in respect to their affinity to living organisms, in terms of behavior, intelligence, and material constitution.

1.4.3 Material Intelligence

> The possibilities in interactive architecture from the vantage point of a biological paradigm make the mechanical paradigm seem dated, ironically before it had a chance to fully manifest itself.
>
> Fox and Kemp (2009, p. 246)

In relation to kinetics and movement Fox and Kemp predict "the end of mechanics" as a "paradigm shift from the mechanical to the biological in terms of adaptation in architecture." This, they argue, will be achieved through novel material developments, especially smart materials, and the close look at biological systems, essentially mimicking natural solutions (Fig. 1.13). They refer to Janine Benyus who defines biomimicry as copying or being inspired by natural designs or processes to solve human-related problems and ensure a more sustainable future (Benyus 1997, p. 2). Benyus distinguishes between three levels of biomimicry:

Fig. 1.13 The smart facade of Cloud 9's Media-TIC building in Barcelona filters solar radiation through responsive ETFE cushions, imitating nature's adaptive qualities, according to the architect Enric Ruiz-Geli (Kretzer 2012)

The first level is based on copying forms or structures that appear in nature, which depending on their usage and context might be more efficient or aesthetic than others. The second level refers to the mimicking of natural processes, of how something is made and produced, and the last part is related to the imitation of natural systems, reasoning that any kind of biomimicry has to be evaluated within a larger context for its long-term purpose and effect (Benyus 2008, p. 40). Petra Gruber, author of *Biomimetics in Architecture*, proposes to focus on the second, processes, and transfer biomimetic principles into the design methods for the creation of innovate architectural solutions. While on a design level a number of novel biomimetic principles, such as self-organization, growth, reaction, homeostasis, or evolution are already implemented for form-finding or simulation, on a structural and material level the amount is rather limited. Yet she believes that great potential lies in "the activation of architectural elements" which "seems to be the most interesting approach that is currently taken towards a living architecture" (Gruber 2011, pp. 7, 190–191).

Condemning biomimetics as "an aesthetic formalism or metaphor for sustainable, but essentially unworkable, aspirations within an urban context," the Co-Director of AVATAR (Advanced Virtual and Technological Architectural Research) in Architecture & Synthetic Biology at the University of Greenwich, Rachel Armstrong calls for the creation of materials and structures that are dynamically connected and responsive to their surrounding. Since the incorporation of real biological systems into built spaces involves energy and maintenance intense support she suggests the invention of a new kind of synthetic biology, native to its surrounding and inherently sustainable, following a bottom-up approach (Armstrong 2011, p. 72). One possible alternative would be so-called Protocells (Fig. 1.14), which Martin Hanczyc from the Laboratory of Artificial Biology at Trento University, describes as "simple chemical models of living cells that possess some of their properties, such as metabolism, movement, replication, information, and evolution but are not necessarily alive." Hanczyc compares the principles of synthetic biology to the creation of architecture, which also (self-) assembles modular building blocks into larger structures that exhibit functionality and form arising from the system as a whole but not the individual elements. To him the benefits of protocell technology are in building the basis for new smart materials applicable to the future of architectural design (Hanczyc 2011, pp. 27–28).

Fig. 1.14 Protocells embedded in Philip Beesley's Hylozoic Ground installation, exhibited in the Canadian Pavilion at the Venice Biennale 2010 (Kretzer 2010)

1.5 Information Materials

> Within the realm of physical materiality, the behavior of responsive materials and assemblies challenges the notion of materials as fixed and dimensionally stable as it highlights the acute propensity for interaction of materials with external energy fields, such as temperature, humidity, sound, and light-as well as to human prompts.
>
> Ng and Patel (2013, Performative Materials in Architecture and Design, p. 5)

As Rashida Ng from the Tyler School of Art at Temple University points out, responsive materials, also often referred to as active, smart, or transformative materials, exhibit great potential for the integration into buildings, and in some cases are already celebrated as the "answer for the twenty first century's technological needs" (Addington and Schodek 2005, p. 1). The architect and materials researcher Blaine Brownell defines such materials as materials that "undergo a physical morphosis based on environmental stimuli." They can be controlled and reversibly alter their properties over time and work both as sensors and actuators (Brownell 2006, p. 120).

In April 2014 a study, called *Envisioning Emerging Technologies*, was conducted by the Envisioning Technology Research Foundation and Policy Horizons Canada, which investigated potential technological advances in six main sectors, chosen for their probability for major disruptive impact on work, life, firms, and politics over the coming 15 years. A particular area was dedicated to "Nanotechnology and Materials Science," which was further divided into three key aspects: Next-generation materials (including active and smart materials that respond to environmental changes), functional materials (that are mostly intended to improve performance and add additional behaviors), and self-assembling materials (which are about making large-scale products more precise). Figure 1.15 displays the three areas and key material developments in relation to their "possible manifestation range" and their estimated maturity (Envisioning Technology Research Foundation 2014). Smart materials, which only represent a fraction of the investigated area, are projected to impact on a larger scale around 2025.

Another report, which was published in January 2015, describes the global market for smart materials at about 26 billion USD in 2014 and expects it to pass 42.2 billion USD by 2019 at an annual growth rate of 10.2 %, with actuators and motors representing the largest application segment of the market (BBC Research 2015).

Mel Schwartz, editor of the *Encyclopedia of Smart Materials* points out that while many smart materials are still immature in regard to practical applications "piezoelectric and electrostrictive ceramics, piezoelectric polymers, and fiber-optic sensor systems" as well as "micromachined electromechanical systems (MEMS), magnetostrictive materials, shape memory alloys (SMA) and polymers, and conductive polymers" are among the more established ones (Schwartz 2002, p. V). Key driving forces for smart material development are the desires for energy saving and the rising popularity of new touch screen

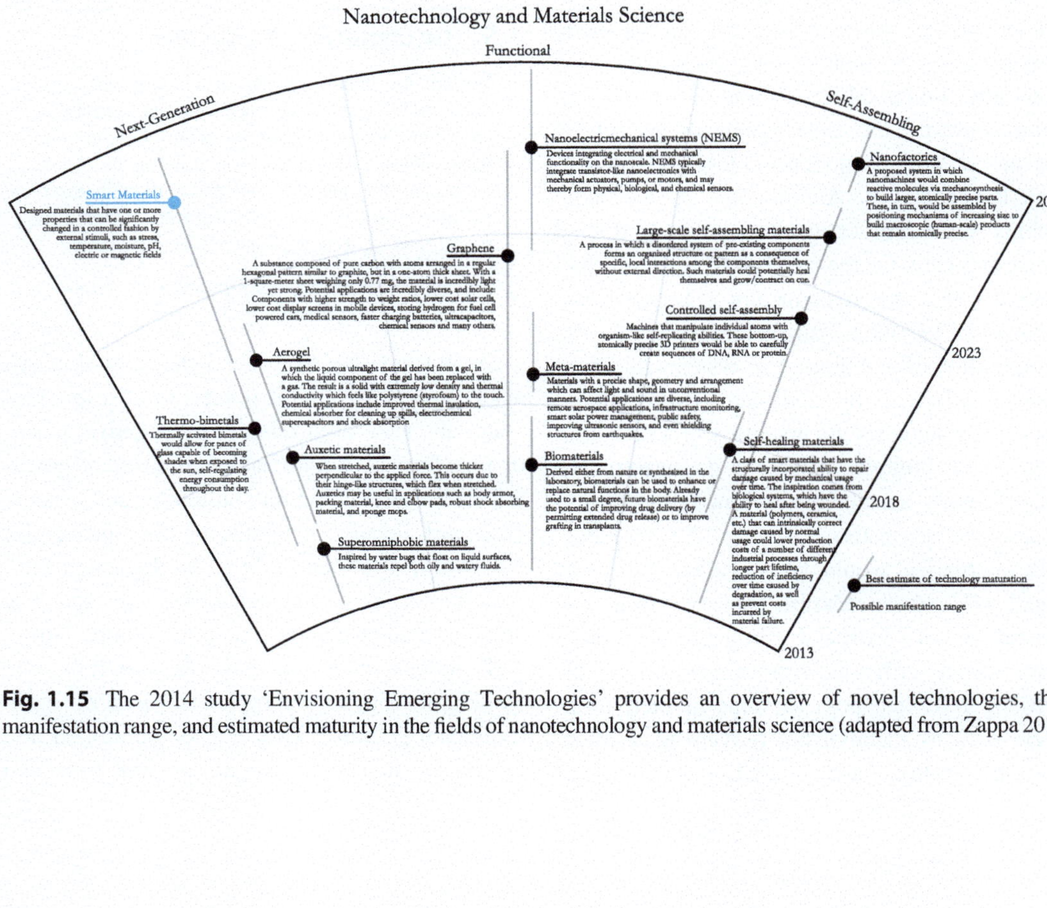

Fig. 1.15 The 2014 study 'Envisioning Emerging Technologies' provides an overview of novel technologies, their manifestation range, and estimated maturity in the fields of nanotechnology and materials science (adapted from Zappa 2014)

devices. Hence, the electronics and automotive industry currently represent the major market for smart material technology, with other areas focusing on medical, marine, mass transit, and aerospace applications (Foresight Smart Materials Taskforce 2003).

When evaluating such reports and projections it becomes obvious that smart materials not only exhibit immense potential for future product development, but also the creation of adaptive architectural spaces. In order to distinguish the approach presented in this book from existing ones the term *information materials* is being introduced. Although it might sound paradox to use yet another name, referring to the same type of materials, there are several reasons that justify the decision.

Information materials aims not at specifying particular material properties but at describing a larger, more general phenomenon, which completely distinguishes such materials from traditional ones, mainly due to their inherently dynamic nature. Yet the focus is not exclusively on materials that possess active and transformative properties but rather tries to capture a current tendency that celebrates the artificial design and creation of materiality exhibiting functionality at a scale formerly impossible. *Information materials* thus not only carry and visualize information but are also based on information, being artificially created from pure intellect. The term hence refers both to the actual materials it embraces and to the synthetic process of their creation. *Information materials*, wants to establish an awareness of a new materiality that can actively change its properties over time, that can be produced and programmed to display a particular performance, and that emerges across multiple disciplines, including nanotechnology, materials science, genetic engineering, or synthetic biology. Most of all, however, it aims at mediating an understanding that these materials are fundamentally distinct from traditional ones. Therefore, they should not only be contemplated in isolation but also lead to profoundly new concepts in terms of architecture and spatial design, attributing value to processes, temporality, and transience. The term *information materials* wants to propose a new type of thinking and establish the necessary educational methods to promote it; thinking that breaks out from a materialistic point of view toward thinking in relation to *materiability*, the ability to create materials with abilities.

1.5.1 Potential of Information Materials for Architectural Usage

In their book *Move*: *Architecture in Motion*, Michael Schumacher et al. very pragmatically see the advantages of such materials "in the potential to make constructions lighter, more slender and more comfortable" and believe that they will "have a significant impact on the appearance and the impression of buildings as well as improve their functionality" (Schumacher et al. 2010, p. 88) Material scientist Lochmatter (2007, p. 6) argues that their main benefits lie in integrated actuator technology:

> In contrast to conventional transducers, such actuators are not mechanically complex since they do not consist of many interacting (passive) parts. Especially when it comes to active structures, the advantages of active materials over conventional actuator technologies [...] become obvious: with conventional technologies one would end up with a very complex framework structure, consisting of interlinked rigid parts [...]. Using active materials, however, a separation between the structure and the driving actuators can be avoided.

Author and editor of the journal *Nature* Philip Ball also highlights the core advantage of *information materials* in simplifying engineering complexity due to their holistic setup. This not only reduces potential hardware complications that could for example occur in mechanical systems but also lowers maintenance requirements and fabrication cost (Ball 1997, p. 104). Other advantages lie in the materials' potential for structural health monitoring, vibration control, and the adjustment of the user environment to increase the performance and energy efficiency of buildings.

Despite the growing interest in the performative aspects of this kind of materiality most spatial scenarios are still rather traditional and fall far from exhibiting any of the radical opportunities *information materials* offer. One highly interesting aspect of many active materials is for example their nonlinear actuation/transformation ratio, which in most cases is regarded as a disadvantage since efficiency and precision are major concerns (Wang 2002, p. 1001). Although controlling them in a predictable and consistent way poses undeniable technological challenges, such a demand in itself speaks for a mechanistic point of view, which in respect to *information materials* is outdated and inappropriate. The materials' transformative processes should instead be seen in a more universal and independent manner, emphasizing their distinction from the rigidity of mechanical paradigms. This would allow focusing on softness and organicism as particular behavioral properties akin to phenomena that occur in nature, which thus might foster a much more intuitive and personal human association than mechanic systems would ever allow for.

1.5.2 Shortcomings of Current Information Material Usage

Next to a lack in ideologically distinguishing *information materials* from traditional ones and a technological immaturity or architectural inadequacy due to limitations in scale or durability, several other reasons so far hinder their successful integration into buildings.

1.5.2.1 Conjunction with Rigid Materials

In many cases, the dynamic properties of *information materials* cannot be used to their full extent as they are either attached to or, as Neri Oxman, assistant professor at the MIT Media Lab, calls it, "patched atop an existing structural or architectural system," which constrains their abilities (Oxman 2010, p. 83). This not only diminishes the potential and efficiency of the respective materials but can also lead to awkwardly forced combinations and assemblies. Such doing is often the result of celebrating the aesthetic effects of the materials for their spectacularity rather than focusing on their performative qualities in terms of responding to human or environmental behavior.

1.5.2.2 Replacement of Existing Technologies

Quite often *information materials* are simply used to replace existing technologies or as Michelle Addington and Daniel Schodek, authors one of the earliest and most influential monographs on smart materials in architecture, put it: "when a new technology has opened the door to unprecedented possibilities [...] architects and designers often try to make it fit within their normative use" (Addington and Schodek 2005, p. 218). A popular example is the light bulb, where the change from incandescence toward LED certainly brought benefits regarding longevity and power consumption, yet insisting on the traditional pear shape is like celebrating "a relic of a former manufacturing age," as CNBC reporter Rosenbaum (2013) observes. Obviously, the more established a certain technology is, the more difficult it is to fundamentally revise it, since it might imply questioning sociocultural values, traditions, and in case of the light bulb even the iconographic representation of ideas.

1.5.2.3 Standardization and Inclusion in Established Catalogs

In many engineering disciplines materials are generally divided into three core categories, which are based on the atomic bonding forces of the particular material: metals, polymers, and ceramics. Other systems distinguish five classes: metals, ceramics and glasses, polymers, composites, and semiconductors (Shackelford 2004, p. 4). In architecture additional types are frequently added based on more visual or tactile properties. Accordingly, following the current attitude toward materials in architecture and in favor of their aesthetic appearance, *information materials* are being evaluated, standardized, and categorized to fit into existing design palettes and catalogs. However by doing so, the active and variable properties of the materials have to be ignored or seriously simplified in order to make them comparable to nonactive materials.

1.5.2.4 Scientific Mystification and Commercial (Un-) Availability

Unfortunately only very little information on new material developments is communicated to the fields of architecture and design in a way that can be comprehended without having expert knowledge or insights. Especially, since architecture is much more related to engineering than it is to materials science, biology, or chemistry, attempting to understand these principles often requires serious commitments and efforts. Moreover, the fact that it usually takes decades until prototypical materials are available as applicable products on the market greatly slows down the creative process and restricts designers to think within established boundaries. On the one hand, this means that they have to rely on commercially available *information materials* that have rather limited properties since they need to comply with building standards and compete with the durabilities and efficiencies of existing technologies. On the other hand, they are restricted to use products that were initially developed for nonarchitectural purposes and contexts and adapt either their design or, in the much lesser cases, the materials to their particular demands.

1.6 Toward a New Softness

> It is change, continuing change, inevitable change, that is the dominant factor in society today. No sensible decision can be made any longer without taking into account not only the world as it is, but the world as it will be. […] This in turn means that our statesmen, our businessmen, our everyman, must take on a science-fictional way of thinking.
>
> Asimov (1981, My Own View, p. 5)

This book is based on the idea that *information materials* have the potential to facilitate a new kind of architecture. In contrast to existing building, which still largely follows the Vitruvian notion of 'firmitatis, utilitatis, venustatis,' stability, usefulness, and beauty, this new architecture will be dynamic, responsive, and to some extent maybe even alive. It will evolve and change with its inhabitants and the environment and instead of being made to endure it will celebrate the ephemeral and temporal aspects of life. Just as humans might turn into technologically enhanced cyborgs,[1] such architecture will merge living, artificial, and digital matter into metaphysical fluid envelopes. Spaces will become unified and holistic **monocoque** systems that will finally allow overcoming outdated mechanistic and structural paradigms, such as the still prevalent separation of a building into 'skin and bones' (Cohen 1996, p. 32). Surfaces will turn into active input/output devices representing and becoming part of a person's hybrid identity. Walls will dissolve and smoothly connect spaces with their related surroundings. Materials will contain and carry information and energy and become active agents in networked ecologies. Architecture will at last evolve from its primordial purpose of providing shelter and safety toward becoming a new synthetic organism, fusing and coexisting with its inhabitants. This new kind of architecture will not only challenge the current way of building in terms of production, processes, and materiality but propose a fundamental ideological rethink, just as Salvador Dali once affronted Le Corbusier by suggesting that the future of architecture will be "soft and hairy" (1957, p. 29).

Clearly within the present context such a vision sounds a lot like science fiction and considering the slow pace of architectural evolution, it might be way too far fetched. Especially, since the function of architecture involves cultural, social, religious, and political aspects among which stability and persistence develop an entirely different meaning. On the other hand the rate at which technology is advancing and becoming more readily available, paired with the shifting demands of upcoming generations toward space and the natural environment represent tendencies that should under no means be overlooked. Considering the unprecedented possibilities and advantages of *information materials* they will thus undoubtedly find their way into architecture, for the time being their

[1]See Sect. 1.2.1 *Technological Advancements* for more details.

revolutionary impact might however be more based on the incorporation of functional applications and products. To elevate such applications to become integral parts of spatial design, rather than just replacements for existing technologies, it will be essential for architects to actively engage in their development at an early stage. Truly interdisciplinary progress can however only happen if all involved parties share and communicate similar interests and possess common knowledge of each other's area of expertise.

Since current approaches toward *information materials* are either too general or conversely too specific for an architect's broad perspective and certainly not valued enough in the present debate new models are imperative. The successful extraction and mediation of such knowledge will hence form an essential requirement for the progressive coalescence of architectural building and *information materials*.

The more immediate and less speculative aspect of this work is thus the assumption that the education of an *information material literacy* will enable architects and designers to abstract and anticipate the functionality, behavior, and usage of any type of (future) *information material*. Moreover, this knowledge will empower them to vitally collaborate across various disciplines and collectively drive the development of an *information material* enhanced architecture of the future.

1.6.1 A Different Kind of Material Thinking

Undoubtedly, architects are largely aware of the value of dynamic and responsive environments in relation to changing occupational demands and user requirements. Likewise they are, and always have been, curious in employing new materials and technologies for addressing such tasks. Yet as mentioned earlier, this often falls short due to a lack of access, experience, and knowledge. The main objective of this book is thus the argument for a different kind of material thinking in relation to *information materials* and architecture, based on two core assumptions.

- First, in contrast to traditional materials, *information materials* are not 'found' and then postprocessed but rather 'created' on a molecular or microscopic level to perform a certain task or to exhibit special properties.
- Secondly *information materials* possess a dynamic immediacy, meaning that they consist of multiple variegated states that change over time and in response to external stimuli.

The aim of this book is to mediate such awareness, focusing on dynamic processes, and behaviors and their deliberately designed artificial constitution, early on during an architect's education. This is ought to empower architects to develop a natural proficiency in anticipating the value and possibilities of *information materials* and apply them in future architectural design. The goal is thus not to propose specific technical solutions or applications, like a catalog explaining how and where to use certain *information materials* in architecture. Rather it aims at discussing their potential and meaning in a more universal and autonomous perspective and provides related knowledge in an open and inspirational manner.

From these assumptions a number of questions arise:

- How has material development and usage evolved in architecture over the course of time and how has this influenced the practice and education of the architect?
- What is the current meaning and usage of *information materials* in architecture and how might it change in the future?
- What are didactic models that focus on the mediation of dynamics, processes, and behaviors and how can explorative material research become a vital aspect of architectural education?

- How can a distinct pedagogic approach be evaluated and how can research results be disseminated to a large, global audience?

1.6.2 Structure of the Book

Information materials are fundamentally distinct from traditional materiality, and hence require a different conception, both theoretically and practically, to make them vital aspects of architectural design. This argument is structured around three interdependent and correlating sections:

- A theoretical part, which elaborates on the continuously changing meaning of materiality throughout architectural history and the role of *information materials* within the present discourse.
- An educational framework, which expands upon existing advances toward *information materials* paired with strategies in accessing and productively applying digital information with the aim of developing an *information material literacy*.
- An evaluation of a selection of *information materials*, both in terms of scientific essays and practical experiments regarding their fabrication procedure and accumulative assembly in speculative architectural installations.

In unison these parts build the foundation for a particular educational method, which regards speculative material exploration as a crucial element of advanced architectural education. The practical aspects of this approach have been tested and validated throughout numerous student courses and workshops. This format allows quickly reviewing and reiterating research results and methodologies within a local and controlled context. Additionally, in order to reach a more substantial mass and prove the concept in a less predetermined environment, an online platform has been established which spreads the acquired information in the form of essays, tutorials, and sample projects to a global network of interested people.

1.6.3 Line of Argumentation

1.6.3.1 Chapter 2: The Ever-Changing Nature of Materiality and the Meaning of Materials in Architecture and Construction

The core of *information materials*, their inherently dynamic constitution, goes far beyond functional and technological aspects and actually questions the very foundation of architecture, its stasis, and permanence. In order to gain a deeper understanding of the implications and possible effects of *information materials* one thus has to reach further than investigating the practical aspects of materials but has to explore the meaning of substance and materiality in a much more theoretical sense. The main focus of this part is to mediate a comprehensive analysis of the notion of materials throughout history by looking at various architectural treatises and on how material culture was understood within the respective context.

The chapter is structured chronologically and divided into five main parts:

The part *Natural Materials* goes from Roman times until the emergence of French Neoclassical architecture at the beginning of the nineteenth century. It passes through the writings of Vitruvius, Villard de Honnecourt, Leon Battista Alberti, Giorgio Vasari, Gian Lorenzo Bernini, Claude Perrault, Marc-Antoine Laugier, Jacques-Francois Blondel, and is concluded with a look at the work of Claude-Nicholas Ledoux.

Industrial Materials highlights some key material developments of the Industrial Revolution such as steel, glass, and reinforced concrete, unfolding in the works of Jean-Nicolas-Louis Durand, Gottfried Semper, Joseph Paxton, and finally, the writings of Le Corbusier who marks the transition to the next part.

Synthetic Materials explores the progress of organic chemistry, which in the mid-twentieth century promised to revolutionize architecture through the formative integration of new types of polymers. It looks at the works of Richard Buckminster Fuller and Frederick Kiesler and is concluded with a number of decisive positions and architectural visions during the so-called 'Plastic Age.'

The part *Digital Materials* evaluates the recent importance of computationally generated form and the analogies between digitally controlled fabrication tools and materiality in the writings of Greg Lynn, Michael Speaks, Bernard Cache, Branko Kolarevic, Fabio Gramazio, Matthias Kohler, Ludger Hovestadt, and Neil Gershenfeld. It finishes with different opinions on the benefits and disadvantages of digital design, digital fabrication, and essentially digital materials.

The final section, *Information Materials*, investigates the potential of a dynamic materiality. It addresses a number of writings on research into programmable matter, a selection of theoretical and practical explorations on the potential of smart materials, and current positions on their long-term value in terms of performance, safety, or energy consumption.

1.6.3.2 Chapter 3: An Attempt for the Education of an Information Material Literacy in Respect to Emerging Materials

The ever-increasing amount of new technologies and materials require new approaches for education and teaching, emphasizing interdisciplinary exchange, and collaboration. This chapter aims at the conveyance of a shared language in respect to *information materials* building upon three interrelated concepts.

The first part stresses the value of physical making and hands-on experimentation, which encourages students to understand and emphasize dynamic material properties and develop (adaptive) architectural solutions in relation to time and space. It elaborates upon artistic views on prototyping, exemplified by the creative practice of Kurt Schwitters, Frederick Kiesler, and Charles and Ray Eames. The section is concluded with a description of the educational methodology of the *materiability* approach and its core structure employed during physical workshops and student projects.

The second section highlights the advantages of open access to information, which not only promotes collective exchange and identification with a shared cause but also encourages the decentralization of knowledge and the democratization from established structures. It builds upon the strategies of Stewart Brand, Yona Friedman, and Bernard Rudofsky and is concluded with an explanation of the *materiability research network*, an online platform, material database, and educational community that provides exclusive access to *information materials* and related content.

The last part speaks about the importance of an *information material literacy*, which marks the outcome of the preceding two methods, the physical experience one gains when working with *information materials* and the limitless access to similar information on further materials. Expanding upon a number of current views on the power of the Internet, social media, and consumer focused digital fabrication tools; the section is finalized with an overview of the social aspects of the *materiability research network*, an online platform that was initiated to make emerging material developments more accessible to the architecture and design communities.

1.6.3.3 Chapter 4: A Selection of Emerging Information Materials, Their Properties, Fabrication, and Application in Speculative Spatial Installations

This chapter focuses entirely on the comprehensive evaluation of a number of different *information materials* and the dissemination of these findings to a larger audience. The examined materials and phenomena include:

- Aerogels, which are ultra light materials that exhibit outstanding thermal conductivity and amazing strength, considering their weight of only 3 mg/cm^3
- Bioluminescence, which is a natural phenomenon concerning the organic production of visible light through varying biochemical reactions
- Bioplastics, which are plastics that are inherently decomposable and largely carbon neutral
- Dye-sensitized solar cells, which are highly efficient, low cost solar cells that use a dye to convert sunlight into energy
- Electroluminescent displays, which are flexible thin-film sheets that emit a perfectly homogeneous cold light
- Electroactive polymers, which are elastomers that change in size and shape in response to a strong electrical field
- Soft Robotics, which is a scientific area that explores the use of pneumatically inflatable silicon actuators
- Thermochromics, which are substances that can change their color reversibly in response to a change in temperature

Each material or phenomenon is described in terms of its historical development and purpose, structure and operating principle, fabrication process, common use, and (potential) architectural application. The fabrication process of every material is reconstructed and demonstrated as detailed and annotated step-by-step instructions. For each part, the intelligibility of the theoretical research and the developed instructions are evaluated within a series of experimental spatial installations that have been realized throughout teaching modules and workshops. This format allows to link theory, material experiments, and environmental studies to a human scale experiential situation.

1.6.3.4 Chapter 5: Toward a New Softness

The final chapter summarizes the work and critically responds to the in Sect. 1.6.1 *A different kind of material thinking* posed questions. It evaluates the initial hypothesis in respect to the value of *information materials* for the creation of adaptive environments and provides an outlook into their potential future.

References

Addington M, Schodek DL (2005) Smart materials and new technologies for the architecture and design professions. Architectural, Oxford

Airbus (2013) The future: the concept plane, Smart Tech Zone. http://www.airbus.com/innovation/future-by-airbus/the-concept-plane/the-airbus-concept-cabin/smart-tech-zone/. Accessed 22 July 2013

Armstrong R (2011) How protocells can make 'Stuff' much more interesting. AD Archit Design 81(2: Protocell Architecture):68–77. doi:10.1002/ad.1209

Asimov I (1981) My own view. In: Asimov on science fiction, 5th edn. Doubleday, Garden City

Ball P (1997) Made to measure: new materials for the 21st century. Princeton University Press, Princeton

BBC Research Market Forecasting (2015) Smart materials and their applications: technologies and global markets. http://www.bccresearch.com/market-research/advanced-materials/smart-materials-applications-technologies-markets-report-avm023e.html. Accessed 5 Mar 2015

Beesley P, Khan O, Stacey M (2013) Acadia 2013: adaptive architecture. In: Proceedings of the 33rd annual conference of the association of computer aided design in architecture. Riverside Architectural Press, Waterloo

Benyus JM (1997) Biomimicry: innovation inspired by nature. Perenniol, New York

Benyus JM (2008) A good place to settle: biomimicry, biophilia, and the return to nature's inspiration to architecture. In: Kellert SR, Heerwagen J, Mador M (eds) Biophilic design: the theory, science, and practice of bringing buildings to life. Wiley, Hoboken, pp 27–48

Brasche S, Bischof W (2005) Daily time spent indoors in German homes—baseline data for the assessment of indoor exposure of German occupants. Int J Hyg Environ Health 208:247–253. doi:10.1016/j.ijheh.2005.03.003

Brownell B (2006) Transmaterial 1. Princeton Architectural Press, New York

Carson R (1964) Silent spring. Fawcett, Crest Books, New York

CBSNEWS (2008) U.N.: World population increasingly urban. http://www.cbsnews.com/news/un-world-population-increasingly-urban. Accessed 10 Mar 2014

Central Intelligence Agency (2013) The World factbook, country comparison: life expectancy at birth. https://

www.cia.gov/library/publications/the-world-factbook/rankorder/2102rank.html. Accessed 22 July 2013
Cohen J-L (1996) Mies Van Der Rohe. Spon, London
Dali S (1957) Dali on modern art: the cuckolds of antiquated modern art. Dial Press, New York
Dezeen (2014) Scary' Venice architecture Biennale show has 'nothing to do with design' says Koolhaas. http://www.dezeen.com/2014/06/05/rem-koolhaas-venice-architecture-biennale-2014. Accessed 5 June 2014
Earth Policy Institute (2014) China leads World to solar power record in 2013. http://www.earth-policy.org/indicators/C47. Accessed 17 Mar 2015
Envisioning Technology Research Foundation and Policy Horizons Canada (2014) Envisioning emerging technologies. http://www.envisioning.io/horizons/. Accessed 5 Mar 2015
Foresight Smart Materials Taskforce (2003) Smart materials for the 21st century. Institute of Materials, Minerals and Mining, Foresight Materials Panel
Fox M, Kemp M (2009) Interactive architecture. Princeton Architectural Press, New York
Frazer J (1995) An evolutionary architecture. Architectural Association, London
Friedman Y (2006) Pro domo. Actar, Barcelona
Gruber P (2011) Biomimetics in architecture: architecture of life and buildings. Springer, Vienna
Hanczyc M (2011) Structure and the synthesis of life. AD Archit Design 81(2: Protocell Architecture):26–33. doi:10.1002/ad.1209
Haque U (2007) Distinguishing concepts: lexicons of interactive art and architecture. AD Archit Design 77 (4: 4dsocial: Interactive Design Environments):24–31. doi:10.1002/ad.487
Hill J (2003) Actions of architecture: architects and creative users. Routledge, London
Hovestadt L (2013) A fantastic genealogy of the printable. In: Hovestadt L, Bühlmann V (eds) Printed physics—metalithikum I. Ambra V, Vienna, pp 17–70
Jaskiewicz T (2013) Towards a methodology for complex adaptive interactive architecture. PhD dissertation, TU Delft
Jeska S (2008) Transparent plastics: design and technology. Birkhäuser, Basel
Kerin L (2010) Alpha generation to be the biggest yet. ABC News. http://www.abc.net.au/am/content/2010/s2851330.htm. Accessed 24 July 2013
Klepeis NE, Nelson WC, Ott WR, Robinson JP, Tsang AM, Switzer P, Behar JV, Hern SC, Engelmann WH (2001) The national human activity pattern survey (NHAPS): a resource for assessing exposure to environmental pollutants. J Eposure Sci Environ Epidemiol 11:231–252
Kohut A, Taylor P, Keeter S, Parker K, Morin R, Cohn D'V, Lopez MH, Smith G, Fry R (2010) Millennials: confident. Connected. Open to change. Pew Research Center, Washington, DC
Koolhaas R (2002) Junkspace. October 100:175–190
Kretzer M (2011) Next generation materials. In: Ambience'11 international conference proceedings. CTF The Swedish School of Textiles, University of Borås, Borås, pp 163–170
Kretzer M (2014a) Architecture in the era of accelerating change. In: Gerber D, Huang A, Sanchez J (eds) ACADIA 2014: design agency. Proceedings of the 34rd annual conference of the association for computer aided design in architecture. University of Southern California, Los Angeles, pp 463–472
Kretzer M (2014b) Materials in the anthropocene. In: Voyatzaki M (ed) What's the matter? Materiality and materialism at the age of computation. European Network of Heads of Schools of Architecture, pp 499–512
Kretzer M (2015) Why so serious? Ehituskunst #57: What do you want, Brick? 42–51
Kronenburg R (2007) Flexible: architecture that responds to change. Laurence King, London
Kruft H-W (1994) A history of architectural theory: from Vitruvius to the present. Zwemmer, London
Kurzweil R (2005) The singularity is near: when humans transcend biology. Viking, New York
Lochmatter P (2007) Development of a shell-like electroactive polymer (EAP) actuator. PhD dissertation, Swiss Federal Institute of Technology (ETH), Zurich
McCrindle M (2008) The ABC of XYZ: generational diversity at work. McCrindle Research Pty Ltd, Cherrybrook
Music Industry News Network (2010) Generation Z reveal expectations for a digital future in global study by habbo hotel. http://www.mi2n.com/press.php3?press_nb=131273. Accessed 24 July 2013
Ng R, Patel S (2013) Performative materials in architecture and design. Intellect, Bristol
Oosterhuis K (2002) Architecture goes wild. 010 Publishers, Rotterdam
Organisation for Economic Co-operation and Development (2012) OECD environmental outlook to 2050: the consequences of inaction. http://www.oecd.org/env/indicators-modelling-outlooks/49846090.pdf. Accessed 15 May 2015
Oxman N (2010) Material-based design computation. PhD dissertation, Massachusetts Institute of Technology
Palfrey J, Gasser U (2008) Born digital: understanding the first generation of digital natives. Basic Books, Philadelphia
Rosenbaum E (2013) How many consumers does it take to design a light bulb? http://www.cnbc.com/id/100910667. Accessed 10 Mar 2014
Schumacher M, Schaeffer O, Vogt M-M, Müller A (2010) Move: architecture in motion—dynamic components and elements. Birkhäuser, Basel
Schwartz M (2002) Encyclopedia of smart materials. Wiley, New York
Shackelford JF (2004) Introduction to materials science for engineers, 6th edn. Macmillan, New York

Toffler A (1970) Future shock. Random House, New York
United Nations (2007) City planning will determine pace of global warming. http://www.un.org/News/Press/docs/2007/gaef3190.doc.htm. Accessed 15 Sept 2010
United Nations, News Centre (2013) World population projected to reach 9.6 billion by 2050—UN report. http://www.un.org/apps/news/story.asp?NewsID=45165#.Ue0wn2QmlqI. Accessed 22 July 2013
Wang ZL (2002) Smart perovskites. In: Schwartz M (ed) Encyclopedia of smart materials. Wiley, New York, pp 992–1013. doi:10.1002/0471216275.esm076
Wiener N (1954) The human use of human beings: cybernetics and society. Doubleday, Garden City

2 The Ever-Changing Nature of Materiality and the Meaning of Materials in Architecture and Construction

2.1 Introduction

> It seems hard to believe that there can be found among things anything of solid body. For the thunderbolt of heaven passes through walled houses, as do shouts and cries; iron grows white hot in the flame, and stones seethe in fierce fire and leap asunder; then too the hardness of gold is relaxed and softened by heat, and the ice of brass yields beneath the flame and melts; warmth and piercing cold ooze through silver, since when we have held cups duly in our hands we have felt both alike, when the dewy moisture of water was poured in from above. So true is it that in things there is seen to be nothing solid.
>
> Titus Lucretius Carus, On the Nature of Things, 59 BC, p. 43

For the Roman philosopher, Lucretius everything in the world is comprised of either matter or void. All that is tangible, that can be perceived with human senses has to exist and is therefore of material character. Since "nothing can be created out of nothing," as he blasphemously observes in his *De Rerum Natura* (On the Nature of Things), all matter has to be made up from elementary bodies, too small to be observed by the human eye. Albeit these bodies are in themselves solid and indivisible, the matter they form is in constant flux, a permanent cycle of becoming and decay. Matter to Lucretius is thus a dynamic entity that continuously changes and behaves throughout space and time, dissolving and reemerging from the fundamental particles it emerges from (Carus and Bailey 1921, pp. 32, 45).

While Lucretius' theory in respect to our contemporary scientific knowledge might appear a bit banal it does highlight a number of aspects, which we nowadays seem to try avoiding by all means. Lucretius addresses decay, ephemerality, and time as elementary and natural properties of all that is. To him nothing that we can perceive is eternal and solid, but will change its properties eventually, either through direct influences, like force or temperature, or indirect environmental impact. In today's highly technologized environment we however tend to treat such aspects like atrocities. We identify good and healthy with clean and spotless and not only want our devices to remain immaculate but also our bodies, environment, and social appearance. Especially in architecture the introduction of biomimicry, bioinspiration, or even living biology, often argued for as a return to nature, a redemption for our egocentric ecological barbarism, tends to become a strange contradiction when considering our repulsion to just the slightest sign of life on our buildings, like the stains of patina or the growth of moss and mold.

This observation, whose further exploration in its cultural and social complexity would go beyond the scope and context of this chapter, gets however even more interesting when dealing with materials of which time and transformation are inherent properties: *Information materials,* as defined in the previous chapter. Professor at MIT Architecture and founding Principal of Kennedy

M. Kretzer, *Information Materials*, DOI 10.1007/978-3-319-35150-6_2

& Violich Architecture Ltd., Sheila Kennedy suggests that the notion of such a vibrant materiality implies a shift in architectural theory from "static material properties to dynamic material behaviors" (Kennedy 2011, p. 118).

But obviously the ideological concept of dynamic matter, constantly behaving and transforming, goes far beyond functional and technological aspects and actually questions the very core of architecture, its stasis. Considering the application and purpose of *information materials* in the context of adaptive architecture thus has to reach further than understanding the practical aspects of materials but has to delve into the meaning of substance and materiality in a much more abstract and theoretical sense.

To Lucretius, who was not an architect but a poet and philosopher, this is quite clear: without materials nothing would exist, especially no architecture, or as Jonathan Hill puts it in his book *Immaterial Architecture*: "Bound to each other, the architectural and the material are considered inseparable" (Hill 2006, p. 2). The role and importance of materiality thus includes much more than structural properties but equally informs a person's experience of a building through its aesthetic, visual, and haptic qualities as well as its associated social, cultural, and historical meaning. Both constructive and ornamental elements represent fundamental parts of a larger assembly, whether obviously appealing or on a more subliminal level. Applying the right materials therefore represents a truly demanding task and requires not only knowledge and experience on the various material properties, but also sensitivity and intuition in anticipating their meaning and value over time; a combination, which paired with a clear understanding and interpretation of the term *appropriate*, might essentially distinguish *good* from *bad* architecture.

The Swiss architect Peter Zumthor describes good architectural design as sensuous (Zumthor et al. 2006, p. 2). "Sense emerges," he writes, "when I succeed in bringing out the specific meanings of certain materials in my buildings, meanings that can only be perceived in just this way in this one building." Despite his realization that materiality is infinite, not only in the amount of available materials but also in the fact that "there are thousand different possibilities in one material alone," he believes that materials can "assume a poetic quality in the context of an architectural object, although only if the architect is able to generate a meaningful situation for them." Zumthor's ultimate goal is the creation of emotional space, free from symbolism and premature meaning (Fig. 2.1). He considers form as secondary and instead focuses for as long as possible on use, structure, and materials, which to him are essential for the creation of atmosphere, the sublime quality of architecture to move, touch, and affect (Zumthor 2006, pp. 11, 25).

Fig. 2.1 The facade of Peter Zumthor's Kunsthaus Bregenz consists of etched glass panels diffusing the incident light and redirecting it to the ceilings in the exhibition areas (Kretzer 2015)

Zumthor's approach reminds strongly of Louis Kahn's philosophy, which describes "architecture [as] the thoughtful making of spaces, [...] the creating of spaces that evoke a feeling of appropriate use." Kahn believes that architecture should express spirituality, which he feels was missing in many of the modernist buildings of his time. In order to accentuate building from pursuing purely formal or utilitarian aspects toward the creation of meaningful spaces he urges to reveal its construction and materiality through a careful consideration of their interdependence in relation to the respective task and location. By defining materials as 'spent light' he highlights the importance of a material's inherent visual and tactile qualities, which, if understood appropriately, will then reveal the suitable construction and form:

> If you think of Brick, you say to Brick, 'What do you want, Brick?' And Brick says to you, 'I like an Arch.' And if you say to Brick, 'Look, arches are expensive, and I can use a concrete lintel over you. What do you think of that, Brick?' Brick says, 'I like an Arch.' And it's important, you see, that you honor the material that you use (Kahn 2003, p. 270).

Just as Kahn feels a lack of spirituality, Frank Lloyd Wright (Fig. 2.2) misses the necessary respect toward materiality. Referring to the 'nature of materials' he demands their proper use and contends that every material has its own significance, potential, and limitations:

> Each different material required a different handling, and each different handling as well as the material itself had new possibilities of use peculiar to the nature of each. Appropriate designs for one material would not be at all appropriate for any other material (Wright 1936, p. 184).

Fig. 2.2 The curved exterior of Frank Lloyd Wright's Guggenheim Museum was built using gun-placed concrete, which is sprayed into formwork instead of being poured (Kretzer 2009)

While the above positions bear obvious similarities they also make clear that just as the definition of architecture varies from architect to architect, the attitude toward materiality does, too. What a material is and what it means is thus tightly tied to subjectivity and personality and cannot satisfyingly be described in an objective or rational manner. Since architectural history is vast and rich, providing countless examples and solutions, Peter Zumthor asserts that much can be learned from the ways materials have been used in the past, both ideologically and practically (Zumthor 2006, p. 22). Following his advice, the present chapter depicts a brief, largely chronological cross section throughout the evolution of architecture with a particular focus on theoretical treatises and the way materials are approached and perceived within the respective contexts. The work tries not to classify, simplify, or value certain materials but instead wants to reveal possibilities and positions. It does not aim for completeness nor does it try to propose or oppress a certain genealogy through its sequential order.[1]

2.1.1 Structure of the Present Chapter

The present chapter is divided into two main parts as graphically displayed in Fig. 2.3. The first treats the topics *Natural Materials*, *Industrial Materials*, and *Synthetic Materials*, while the second addresses the areas *Digital Materials* and *Information Materials*.

The topic *Natural Materials* commences from an analysis of the writings of the Roman architect Vitruvius, the first architect to have left written records of his field. It is followed by the Gothic period, which is described through the drawings and sketches published in the early thirteenth century by Villard de Honnecourt. This evaluation is superseded by the Renaissance writings of Leon Battista Alberti and Giorgio Vasari. Addressing the founder of the Baroque style Gian Lorenzo Bernini, the section leaps into the era of Enlightenment, with a particular focus on Claude Perrault, Marc-Antoine Laugier, and Jacques-Francois Blondel. It is concluded with a look at the designs of Nicholas Ledoux, one of the earliest advocates of French Neoclassical architecture.

Part two, *Industrial Materials*, focuses mostly on the effects of the Industrial Revolution on

[1] Ákos Moravánszky argues in his anthology *Architekturtheorie im 20. Jahrhundert* that a chronological "history of architectural theory" impedes a critical and vivid discussion between varying positions, which are not only spatially but also timewise distinct. Hanno-Walter Kruft on the other hand contends that a sequential order allows for the maintenance of a historical continuum, which has narrative advantages and provides stability and ease of orientation.

NATURAL MATERIALS

Vitruvius
De Architectura Libri Decem (27BC)

Villard de Honnecourt
(1230)

Leon Battista Alberti
De re Aedificatoria (1452)

Claude Perrault
Ordonnnance des Cinq Espèces de Colonnes selon la Méthode des Anciens (1683)

Marc-Antoine Laugier
Essay sur l'Architecture (1755)

Jaques-Francois Blondel
Essay sur l'Architecture (1771)

Claude Nicolas Ledoux
L'Architecture Considérée sous le Rapport de l'Art, des Moeurs et de la Législation (1804)

INDUSTRIAL MATERIALS

Jean-Nicolas-Louis Durand
Précis des Leçons d'Architecture données à l'École Royale Polytechnique (1805)

Gottfried Semper
Die Vier Elemente der Baukunst (1850)
Der Stil in den Technischen und Tektonischen Künsten (1878)

Joseph Paxton
Crystal Palace (1851)

Le Corbusier
Vers une Architecture (1923)
Five Points Towards a New Architecture (1926)
Quand les Cathedrales etaient Blanches (1937)

SYNTHETIC MATERIALS

Richard Buckminster-Fuller
Nine Chains to the Moon (1938)
Operating Manual for Spaceship Earth (1968)

Frederick Kiesler
Vitalbau-Raumstadt-Funktionelle Architektur (1925)

Victor Yarsley and Edward Couzens
Plastics (1945)

Monsanto Chemical Company
Monsanto House of the Future (1956)

Yona Friedman
Groupe d'Études d'Architecture Mobile (1956)

DIGITAL MATERIALS

Greg Lynn
Folding in Architecture (1993)

Bernard Cache
Earth Moves: The Furnishing of Territories (1995)

Branko Kolarevic
Architecture in the Digital Age: Design and Manufacturing (2005)

Fabio Gramazio, Matthias Kohler
Digital Materiality in Architecture (2008)

Ludger Hovestadt
Beyond the Grid: Architecture and Information Technology (2009)

Neil Gershenfeld
How to Make Almost Anything: The Digital Fabrication Revolution (2012)

INFORMATION MATERIALS

Tommaso Toffoli and Norman Margolus
Programmable Matter: Concepts and Realization (1990)

Seth C. Goldstein and Todd C. Mowry
Claytronics: An Instance of Programmable Matter (2004)

U. S. Army Research Office Workshop
Smart Materials, Structures and Mathematical Issues (1988)

Foresight, Institute of Materials, Minerals and Mining
Smart Materials for the 21st Century (2003)

Michelle Addington and Daniel Schodek
Smart Materials and new Technologies: For Architecture and Design Professions (2005)

Vera Bühlmann and Ludger Hovestadt
Printed Physics: Metalithikum I (2013)

Fig. 2.3 Overview of literature that has been reviewed, separated into the five parts natural materials, industrial materials, synthetic materials, digital materials, and information materials, which can be sorted according to two main areas (Kretzer 2015)

both architectural building and thinking. Highlighting a number of outstanding material developments, such as steel, glass, and reinforced concrete it unfolds the works of Jean-Nicolas-Louis Durand, Gottfried Semper, and Joseph Paxton and their effects on the profession of the architect and the structures that were designed accordingly. It is concluded with the writings of Le Corbusier who marks the transition to the next episode.

The following section, *Synthetic Materials*, explores advancements in organic chemistry, which during the mid-twentieth century lead to the introduction of new types of polymers that promised to define a new domestic lifestyle. The works of Buckminster Fuller and Frederick Kiesler are explained and the Monsanto House of the Future is described as one of the most publicly promoted experiments into a new type of plastic dwelling. The part is concluded with a number of positions emerging during the 1960s and 1970s and a shift in social thinking that marked the end of the plastic euphoria.

The part *Digital Materials* begins with the writings of Greg Lynn, Michael Speaks, and Bernard Cache on computer-aided design and manufacturing at the turn of the twenty-first century. It continues with an investigation into the meaning of digital materiality, referring to the writings of Fabio Gramazio and Matthias Kohler. Building upon Ludger Hovestadt's discussion on the relationship between advancements in architectural production and the history of biology the section carries on with the development of computer-aided manufacturing technologies and is concluded with Neil Gershenfeld's take on the future of digital fabrication. After highlighting a few path breaking projects that employed digital fabrication technologies at a building scale it concludes with different positions on the benefits and downsides of digital design, digital fabrication, and essentially digital materials.

The last section, *Information Materials*, commences from different scientific approaches toward the creation of programmable matter, in the form of fine-grained particles that can in unison assemble into larger functional elements. The part then turns toward the area of smart materials, describing their history, evolution, and varying terminologies in detail based on a number of related reports and surveys. It highlights their theoretical and practical usage in the area of architecture and construction and describes a number of realized architectural works. The section is concluded with more abstract and theoretical views on the abilities of *information materials* and their related production techniques.

2.2 Natural Materials

> Although it would next be in order to explain the proper proportions and symmetry of temples and public buildings, as well as of private houses, I thought best to postpone this until after I had treated the practical merits of the materials out of which, when they are brought together, buildings are constructed with due regard to the proper kind of material for each part, and until I had shown of what natural elements those materials are composed. But before beginning to explain their natural properties, I will prefix the motives which originally gave rise to buildings and the development of inventions in this field.
>
> Vitruvius Pollio (1486) On Architecture, pp. 35–36

Vitruvius' treatise *De Architectura Libri Decem* (The Ten Books on Architecture), which marks the to date oldest architectural discourse and only surviving record on classical architecture was most likely written between 27 and 23 BC (Kruft, p. 447). The comprehensive work is divided into ten consecutive volumes, each dealing with a certain aspect of Roman building and construction. While the first volume introduces "the functions of architecture and the scope of the art" as well as an architect's ideal training and "what the qualities of an architect should be," the second volume is solely dedicated to "the use of the building materials which nature provides," which are brick, sand, lime, pozzolana, stone, and timber. In order to explain why these are the only relevant materials for Roman building and to convey how to apply them successfully, Vitruvius elaborates on "the origin of the building art, how it was fostered, and how it made progress, step by step, until it reached its present perfection" (Vitruvius Pollio 1914, pp. 35, 41).

De Architectura begins with the assumption that it "was the discovery of fire that gave rise to the coming together of men," an accidental intervention of nature, which lead to earliest social systems, collaborative exchange, and essentially the requirement for shelter. Vitruvius describes the ability of humans to converse and teach one another, which distinguishes them from the rest of living creatures, as the essential key toward improving and progressing structures and methods. He argues further that the earliest human dwellings were either entirely of natural origin, such as caves or holes, or based on the imitation of procedures found in the animal and plant kingdom, such as copying the nest building techniques of swallows.

Disregarding the role of man in the creation of shelter he describes two basic models as the archetypes of human construction, both emerging as logical consequences from the direct use of the respective materials. The first one, as depicted in the French translation of Vitruvius by Claude Perrault from 1673 (Fig. 2.4 Left), is explained as a kind of tower, constructed by placing wooden beams, alternatingly and perpendicular on top of each other. It has a pyramidal roof and "the interstices, which are left on account of the thickness of the building material, are stopped up with chips and mud." The second model, which originated in areas where timber was scarce, is pictured in Fig. 2.4 Right as "a pyramidal roof of logs fastened together, and [covered] with reeds and brushwood."

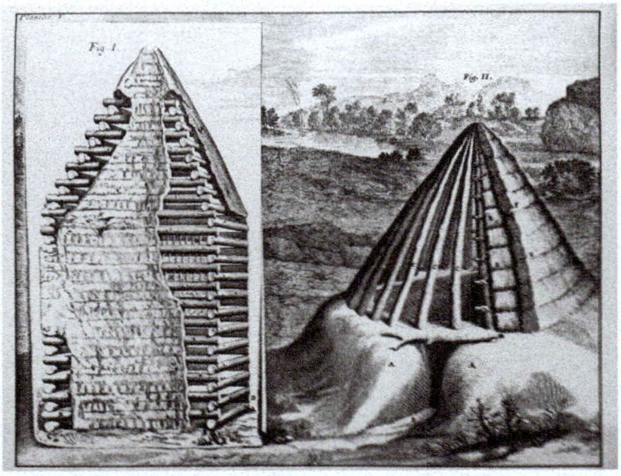

Fig. 2.4 *Left* Graphical depiction of the primitive house of the Colchians; *Right* Drawing of the primitive house of the Phrygians (both from Les Dix livres d'architecture de Vitruve, Claude Perrault, 2nd ed., Paris 1684)

Vitruvius' descriptions are less funded in real archaeological findings than in personal observations he made among the building techniques of various foreign tribes of his time. From these studies he concludes that the evolution of architecture is a logical consequence from the resources that are available in nature as well as the progress man made "by becoming daily more expert in building." Describing the advent of the craft of carpentry as the turning point, when humans "passed from a rude and barbarous mode of life to civilization and refinement," he explains that together with "the multiplication of the arts they gave up huts and began to build houses with foundations, having brick or stone walls, and roofs of timber and tiles." Further intellectual development leads to the understanding and use of symmetry, a focus on style and luxury, and finally culminated in the Greek temple (Fig. 2.5), which to Vitruvius marks the unsurpassed archetype of architectural proficiency (Vitruvius Pollio 1914, pp. 38–40).

Fig. 2.5 The Greek Temple of Segesta, Sicily is thought to have been built in the 420s BC. It follows a Doric order and has six by fourteen columns on a base measuring 21 by 56 m, on a platform three steps high (Kretzer 2015)

2.2.1 The Influence of Vitruvian Thought on Current New Materialism Positions

Vitruvius is however not only impressed by the clarity and elegance of Greek building but also deeply moved by their thinking and philosophy. Referring to Thales, Heraclitus, Democritus, and the school of Pythagoras he asserts, similar to his contemporary Lucretius, that "there is no kind of material, no body, and no thing that can be produced or conceived of, which is not made up of elementary particles," which are water, fire, earth, and air. Since these elements cannot be dissolved, cut, or harmed, they must be eternal and retain infinite solidity. From this he assumes that every material has some inherent natural characteristics that determine the possible structures it allows for and thus the forms that are possible to realize. The Greek temple hence is to him not only the peak of architectural evolution but also presents a logical incarnation of naturalness and material usage and thus must be seen as a direct descendant of the earliest human form of shelter (Vitruvius Pollio 1914, pp. 41, 109).

Similar theories and conclusion have resurfaced multiple times throughout history and gain considerable attention among current 'New Materialism' positions. The philosopher Manuel De Landa describes New Materialism as "based on the idea that matter has morphogenetic capacities of its own and does not need to be commanded into generating form" (Dolphijn and van der Tuin 2012, p. 43).

> We may now be in a position to think about the origin of form and structure [...] as something that may come from within the materials, a form that we tease out of those materials as we allow them to have their say in the structures we create (DeLanda 2004, p. 21).

Rather than considering form or design as pure thought, imposed upon homogenized materiality he suggests the opposite, to treat materials as "active participants in the genesis of form," indicating the existence of a heterogeneous materiality with variable properties (DeLanda 2001, p. 132). Relating this concept to certain principles that can be observed in nature he contends that architectural form finding could benefit strongly from considering the "combinatorial productivity of natural forms" such as, for example, the interplay of "bones bearing loads in compression and muscles bearing them in tension" (Braham et al. 2007, p. 391). Neri Oxman, who heads the Mediated Matter Group at MIT's Media Lab, builds her research upon a similar argumentation:

> As in Nature, when creation begins with matter, morphogenesis, or the generation of form, is a process engendered by the physical forces of Nature. [. .] Material behavior in Nature appears to be a prerequisite for the emergence of form, and yet in design, shape eternally comes first (Oxman 2010a, b, pp. 70–71).

As an alternative she proposes "material based design computation," emphasizing a "nonhierarchical association between form, structure, and material" (Oxman 2010a, b, p. 72). Oxman believes that natural forms, "which [are] directly informed by the materials from which they are made," are more efficient and less wasteful than any of mankind's own material strategies and are thus inherently sustainable (Oxman 2010a, b, pp. 80–81). She adds that even though novel design and fabrication techniques allow for increased variety and complexity, they only support the usage of materials with homogeneous properties, an approach that she considers as outdated in respect to the extremely diverse context of our time. In accordance to biological principles she alternatively aims to "facilitate the variation of material properties through a heterogeneous materiality as a function of environmental performance which thus becomes an integral part of the form-generation process."

In a similar but more structurally determined way demands Achim Menges an interrelated understanding of form, material, and structure based on computational techniques to analyze material capacities, geometrical restrictions, manufacturing processes, and assembly logic:

> In contrast to the integral processes of material formation in nature, architecture as a material practice is still predominantly based on design approaches that struggle to fully explore the materials' richness of performative capacity and resourcefulness for design (Menges 2012, p. 17).

Criticizing contemporary CAD tools that foster a hierarchical relationship between form and material, basically treating materiality as a passive entity to manifest geometry, he proposes a design in which a material's inherent characteristics are utilized as 'morphogenetic drivers' for form generation. Through understanding a material's microstructure in relation to its external appearance as a continuously transforming entity and by applying computation to navigate within this environment he recommends an "exploratory design process of unfolding material-specific gestalt and related performative capacity" (Menges 2012, p. 36).

Both Menges and Oxman share strong affinities in proposing computational form-finding concepts paired with an understanding of materials as heterogeneous entities to oppose the predominant preference of form over materiality in current architectural design. Menges analyzes natural phenomena, such as the tissue of bones or insect shells, as well as the microstructures of existing materials, and then adapts the observed biological principles in junction with material-specific criteria for the creation of experimental pavilions. Oxman argues to follow natural growth principles and in many of her projects utilizes 3D printing or other additive fabrication techniques to build up 'new' materials based on such observations.

Returning to Vitruvius it is interesting to observe that his elucidation, which originates from an anthropogenic technological point of view, by explaining the primordial substances that make up all materials, is commenced with a chapter on brick—the first man-made composite, while retaining the notion that it is a natural material. To him naturalness thus refers to a shared quality inherent in all things in existence, including artificially produced materials and structures. Taking this notion as point of departure he arrives at yet another synthetic material to which he dedicates even more attention, pozzolana concrete:

> There is also a kind of powder from natural causes produces astonishing results. [...] This substance, when mixed with lime and rubble, not only lends strength to buildings of other kinds, but even when piers of it are constructed in the sea, they set hard under water (Vitruvius Pollio, pp. 46–47).

The powder Vitruvius addresses is basically sandy volcanic ash, originally discovered at Pozzuoli, a region around Mount Vesuvius in Italy. Mixed with lime it results in hydraulic cement that can be used to create a strong mortar that even solidifies under water. The discovery of concrete enabled arches, domes, and vaults to be erected without the constraints of masonry and allowed Roman architecture to achieve a substantial leap in construction. The Pantheon, which was built from 118 to 126 AD, holds a hemispherical concrete dome that spans the circular interior of over 44 m. Its building technique and material were so advanced that it remained the largest single building span until the nineteenth century.

2.2.2 The Growing Detachment of Form from Materiality

With the fall of the Roman Empire, Western European architecture experienced a serious regression and a return to mostly wooden materials that lasted for several hundred years. Between the sixth and tenth century Romanesque architecture appeared, which aimed for a reproduction of Roman vaulted styles. The buildings, which had clearly defined, often symmetrical, forms were meant to express wealth and power and were thus made from heavy materials, mostly stone and brick. In comparison to Gothic architecture, which followed in the twelfth century, Romanesque buildings were simplistic and reduced. Gothic architecture originated in France and spread with surprising speed across most of Europe. The term Gothic was however only introduced much later by the Italian painter and architect Giorgio Vasari in his book *Le Vite de' più eccellenti pittori, scultori, e architettori da Cimabue insino a' tempi nostri* (Lives of the Most Excellent Painters, Sculptors, and Architects, from Cimabue to Our Times), published in 1550, as a clear distinction from Renaissance architecture. Vasari describes the period as being "monstrous and barbarous, and lacking everything that can be called order" associating it with the Gothic tribes who fought the Roman Empire from the fifth to eighth century and destroyed the purity of the Roman style (Vasari et al. 1960, p. 83). Quite the contrary the Gothic style was based on sophisticated structural systems and came from the invention of a much lighter and defined type of masonry vault, which in comparison to the massive Roman vaults, consisted of an intricate network of arches, or ribs, that span the space. Little information on Gothic building and design has survived to date, partly because paper was not yet invented and the expensive parchment was only used in exceptional cases (Ackerman 1997, p. 42). Of particular importance is thus a set of architectural drawings and sketches published in the early thirteenth century by the French artist Villard de Honnecourt (Fig. 2.6). To Honnecourt the stylistic core of Gothic architecture are the linear articulating members, both animating the surface and describing the structure (Kidson et al. 1990, p. 136). While the plan of Gothic churches is similar to the Romanesque plan, the interior is much higher and completely surrounded with colored windows allowing light to triumph over substance. But especially on the outside the difference to the earlier periods can be felt, with Gothic exteriors being extremely complex, detailed, and ornamented, including carved figures, mystical rainspouts, called gargoyles, and often a huge round rose window above the large entrance.

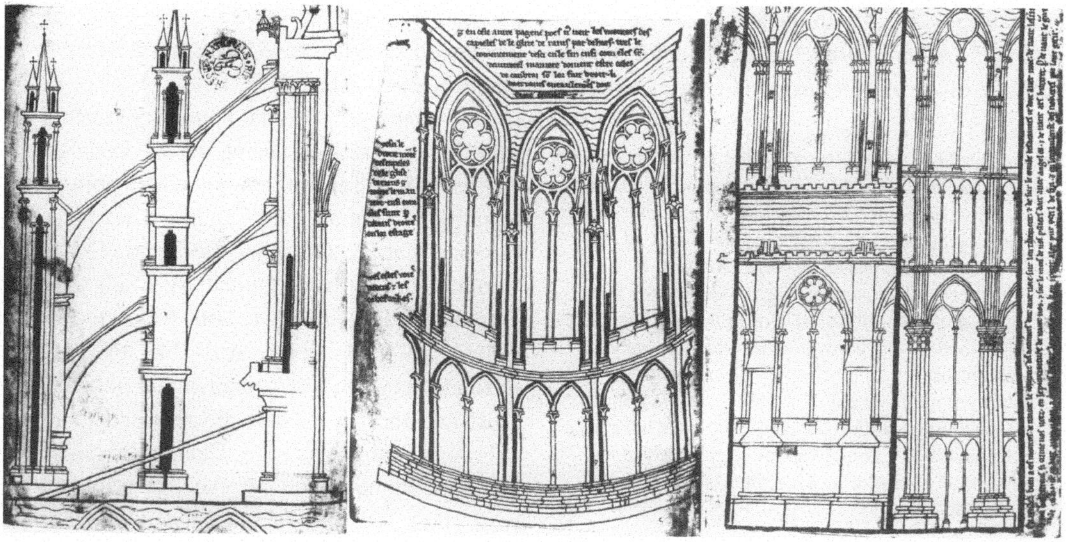

Fig. 2.6 *Left* Reims Cathedral, buttress elevation and nave section; *Middle* Reims Cathedral, interior of choir; *Right* Reims Cathedral, exterior and interior elevation (both from *Sketchbook of Villard de Honnecourt*, Villard de Honnecourt 1220–35)

The Gothic style dominated much of European building, especially cathedrals and churches, until the sixteenth century, except for Italy. Italian architects, such as Vasari, rejected the structural basis of Gothic building and instead sought to revive the spatial magnificence of the Roman era. While Gothic builders created light-filled interior space through complex vaulting, Renaissance architects aimed for symmetry through mathematical proportions. Blaming the 'Goths' for ruining the ancient style and culture Vasari demands to "protect every country from such ideas and style of buildings" which "are such deformities in comparison with the beauty of our buildings" (Vasari et al., p. 83).

A similar, yet not as polemic, position can be found in Leon Battista Alberti's *De re Aedificatoria* (On the Art of Building), published in 1452 during the beginning of the Italian Renaissance. Alberti's work is the first book on architecture since Vitruvius' and just as *De Architectura Libri Decem* is split into ten volumes addressing the various aspects of architectural building. Appearing to be both a tribute and a challenge to the preceding masterpiece the major difference lies in the fact that while Vitruvius mainly describes how things are built in the past and present, Alberti sets out to propose how architecture should be done in the future. Alberti agrees upon Vitruvius' claim that in order to construct a successful building one "should [...] copy the ingenuity of nature," yet his notion of imitation is fundamentally different from the ancient understanding. Whereas Vitruvius believes in some kind of underlying God-given order, to Alberti copying nature means to discover and understand natural principles and then, through a certain intellectual abstraction, apply these rules to works of artistic production. This means that materials, despite their varying qualities, each suited differently well for particular purposes, become subordinate to design and take the role of transforming abstract ideas into physical artifacts.

To Alberti architecture is hence purely intellectual work, performed by the architect and preceding the actual construction of the building, which is possible "without any recourse to the material" (Alberti 1988, pp. x, 7, 86). In the prologue of *De re Aedificatoria* Alberti writes:

> Before I go any farther, however, I should explain exactly whom I mean by an architect; for it is no carpenter that I would have you compare to the greatest exponents of other disciplines: the carpenter is but an instrument in the hands of the architect. Him I consider the architect, who by sure and wonderful reason and method, knows both how to devise through his own mind and energy, and to realize by construction, whatever can be most beautifully fitted out for the noble needs of man, by the movement of weights and the joining and massing of bodies. To do this he must have an understanding of knowledge of all the highest and most noble disciplines. This then is the architect (Alberti 1988, p. 3).

In terms of building materials Alberti is basically content with Vitruvius, describing "lime, sand, stone, timber; and likewise iron, bronze, lead, glass, and so on" (Alberti 1988, p. 38). What is surprising however is the fact that neither Alberti, Palladio, Vasari, nor any other Renaissance writer dedicates any particular significance to pozzolana concrete, the material which Vitruvius describes so extensively.[2] Siegfried Giedion suggests that "concrete disappeared as Gothic architecture continued to pare down the size of its structural members to achieve pure skeletal forms, for concrete had been associated with the massive Roman wall" (Giedion 1971, p. 260). Yet the reason why the material did not reemerge during the Renaissance period, which aimed to both demarcate itself from the Gothic style and return to classical Roman building, remains unclear.

2.2.3 The Domination of Human Intellect Over Natural Materiality

The Renaissance period was superseded by the Baroque era, which began in the late sixteenth century. Strongly promoted by the Catholic Church the new style was based on earlier Roman and Renaissance forms but differed in an unprecedented intensity in terms of mass, color, shadow, and light. Often credited for creating the Baroque style and also being the last Renaissance architect, in terms of having equal abilities in painting, sculpture, and architecture, is Gian Lorenzo Bernini (Boucher 1998, pp. 134–142). Bernini was in 1665 invited by Louis XIV of France, le Roi-Soleil, to finish work on the Louvre in Paris. However, due to growing misunderstanding and disagreement between the artist and the royal court, he returned to Rome only 4 months later and Claude Perrault was instead commissioned for the execution of the work. Perrault's design had only little that could be called Baroque but was an expression of his own understanding of beauty and proportion, which he concluded from a careful analysis of classical architectural treatises.

In the *Ordonnance des Cinq Espèces de Colonnes selon la Méthode des Anciens* (Ordonnance for the Five Kinds of Columns after the Method of the Ancients), written in 1683 Perrault argues:

> All those who have written about architecture contradict one another, with the result that in the ruins of ancient buildings and among the great number of architects who have dealt with the proportions of the orders, one can find agreement neither between any two buildings nor between any two authors, since none has followed the same rules (Perrault 1993, p. 48).

From this variance he reasons that beauty is less attached to proportions and thus an objective quality of buildings, as has been assumed earlier, but more related to subjective interpretation of the viewer and as such dependent on custom, time, taste, and fashion. As society evolves this perception changes, which explains why the proportions and styles of columns have changed accordingly. Perrault neglects that architecture should be an imitation of nature, since he finds no connection in neither the proportions of the human body to the column, nor in its resemblance to a tree trunk. In that sense he is the first to break the classical understanding that the Greek temple is a direct descendant from the primitive hut. Yet

[2] Vitruvius' chapter on pozzolana concrete counts 840 words while his descriptions of brick (650 words), sand (350 words), lime (430 words), and stone (680 words) are much briefer.

Perrault believes that while certain aspects of beauty arise in the eye of the beholder and are thus constructed reality, other, more general, qualities can be measured empirically and evaluated scientifically. The beauty of the Pantheon, for example, is to him not a result of "the proportion of that temple's wall thickness to its interior void," as "most architects claim," since these are unperceivable qualities. Rather it is based on beauty, which he calls of "convincing reasons [...] whose presence in works is bound to please everyone." Perrault's lucid position made him one of the early proponents of the Enlightenment, an intellectual movement, which aimed to break out of tradition and instead follow reason and scientific thought (Perrault 1993, pp. 49–50).

Another highly popular writer of the Enlightenment was Marc-Antoine Laugier who in 1753 published his *Essay sur l'Architecture* (An Essay on Architecture), arguing for a reform of architecture built upon the notion of the primitive hut, a vivid idealization of Vitruvius' theory. The frontispiece (Fig. 2.7) of the second edition depicts a female personification of architecture, directing the attention of a young child toward a simplistic structure consisting of four trees that are holding a triangular roof. In contrast to the earlier Perrault, the naturally grown temple becomes in Laugier's view not only the genesis but also the embodiment of all architecture, representing an immediate imitation of natural processes:

> The little rustic cabin that I have just described, is the model upon which all the magnificences of architecture have been imagined, it is in coming near in the execution of the simplicity of this first model, that we avoid all essential defects, that we lay hold on true perfection (Laugier 1977, pp. 11–12).

Fig. 2.7 An idealized version of the 'primitive hut,' envisioned by Laugier as the epitome and model for architecture (Frontispiece of Marc-Antoine Laugier: Essai sur l'Architecture 2nd ed., Charles Eisen 1755)

Obviously Laugier's model is, even less than Vitruvius' description, not based on archaeological facts but rather a deliberate theoretical position elevating the Greek temple as the true and only origin of building and the model after which new work should be created. Through this highly polemic demand he criticizes the formal excesses of the earlier Baroque and Rococo era and proclaims the return to the fundamental principles of architecture. While Laugier's hut is strongly built upon the idea of natural and divine proportion, found in Vitruvius' writings, it differs in the relationship of man and nature and the role materials play in connecting the two. With Vitruvius not attributing much credit to man as the creator of form but rather arguing that form and structure emerge from immanent material qualities, Laugier's position is that, albeit form should be based on natural principles, it is man who induces the transformation and thus marks the dominant force (Braham 1980, pp. 48–49).

Laugier's Essay had vast and immediate influence on many thinkers of his time and also informed the writing of Jacques-Francois Blondel, who in a very didactic manner proposes in his *Cours d'Architecture ou Traité de la Decoration, Distribution & Construction des Bâtiments* (Course of Architecture), published in 1771, a rational approach toward architecture and building. Just as Perrault, Blondel believes that good architecture is simply a manner of taste, something that cannot be determined, yet there are rules, which can lead to harmony between a building's function, appearance, and structure

(Evers and Thoenes 2003, p. 296). Blondel's work is a demand for order and the development of a 'true style,' which "marks the entry of the modern concept of style into architectural theory," as Hanno-Walter Kruft observes (Kruft 1994, pp. 148–149).

One of Blondel's students was Claude-Nicolas Ledoux who became one of the earliest advocates of French Neoclassical architecture, a style in which classical forms, derived from Vitruvian principles and especially the work of Palladio, were assembled more for their dramatic effect than to create orderly arrangements. Strongly demarcating himself from other architects of his time, Ledoux thinks that architecture needs to propose a better future for society and has to emerge from the imagination of the architect and not rationality and consent (Pérez-Gómez 2005, p. 46).

His *Treatise L'Architecture Considérée sous le Rapport de l'Art, des Moeurs et de la Legislation* (Architecture in its Relations to Art, Customs and Legislation) was published in 1804 and aims to establish a generalized architectural system, including all the various tasks of an architect, thus reinterpreting his role within a visionary social structure. In this system of social coexistence, the architect becomes an educator, with equal political, moral, and religious responsibilities. Similarly radical are his architectural designs, which are massive symbolic monuments, like cubes, spheres, and cylinders, representing strong simplifications of classical building types. In his project for the salt-producing town of Chaux in the French Jura, of which the major part was built between 1774 and 1779, he realized not only some very modern ideas about industrial production, but also designed a series of geometrical monuments that visually express the activities its' tenants are involved in. The hoop makers were thus to live in houses shaped like wheels, the 'Pacifère,' the peacemaker, in a building shielded by fasces that symbolize unity, and the house of the river inspector should have the river running right through it (Evers and Thoenes, p. 320). The most popular of his designs, albeit never realized, is the house of the forestry guard (Fig. 2.8), a spherical building without windows, completely deprived from any functionality or materiality, and only focusing on its pure form and symbolic character.

Fig. 2.8 Ledoux's spherical and windowless 'House of the forestry guard in château de Mauperthuis' was a deliberately geometrical concept to contrast architecture of the time (Claude-Nicolas Ledoux 1789)

2.3 Industrial Materials

> Forms and proportions may be divided into three categories: those that spring from the nature of materials, and from the uses of the things they serve to build; those that custom in a sense made necessary to us, such as the forms and proportions of the buildings of antiquity; and finally, those simpler and more definite forms and proportions that earn our preference through the ease with which we apprehend them. Of these, only those in the first category are essential; but they are not so firmly defined by the nature of things that we cannot add to them or subtract from them, so that there is no reason not to combine them with those of the second class, derived from ancient buildings. Since these vary considerably [...] one is at liberty to select the simplest, [...] the best suited to satisfy both the eye and the mind.
>
> Durand (1831a, b, Précis des Leçons d'Architecture, pp. 108–109)

A major turning point, not only in architecture but essentially every aspect of life, occurred during the second half of the eighteenth century, with the emergence of the Industrial Revolution, first in Great Britain and then, within a few decades, in Western Europe and the United States.

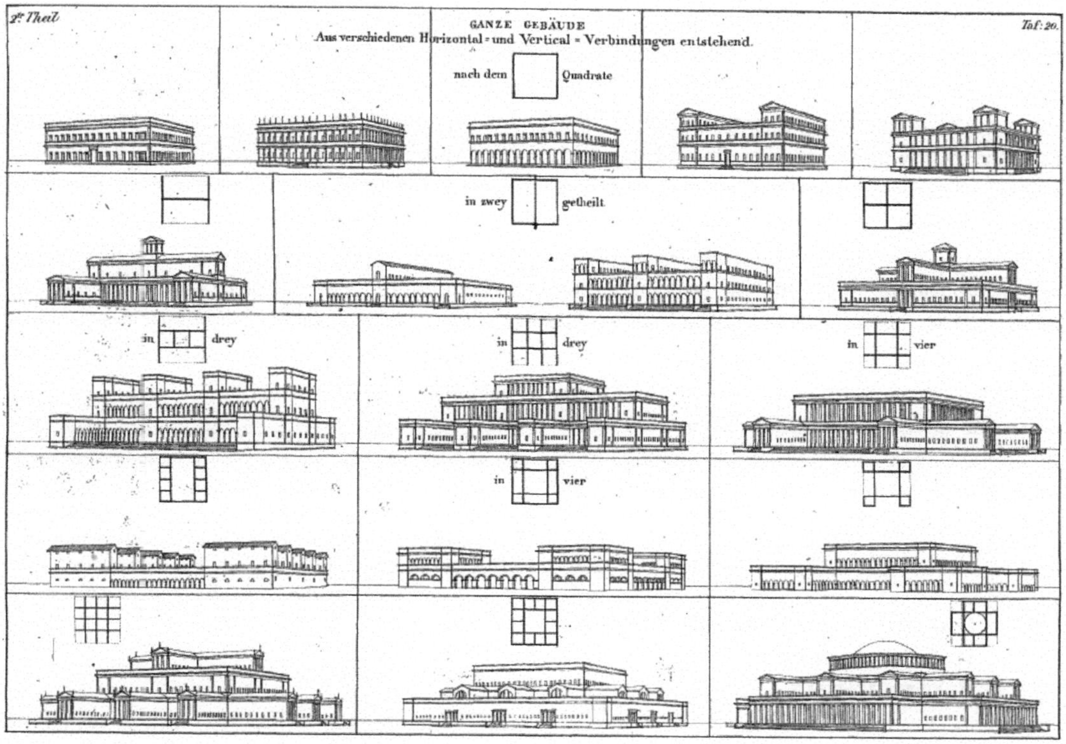

Fig. 2.9 Proposal for buildings based on simple *rectangular* shapes, divided vertically and horizontally (Durand 1831a, b)

One notable architect, who was born into the vast technological transition of automated manufacturing and the mass production of new materials, was the French, Jean-Nicolas-Louis Durand. In 1831 a collection of his lectures at the École Polytechnique in Paris was published, the *Précis des Leçons d'Architecture données à l'École Royale Polytechnique* (Precis of the Lectures on Architecture). The comprehensive work was initially intended as a primer or construction manual for his students but soon became one of the most influential textbooks of his time: a new grammar for architecture. Strongly motivated by the unprecedented possibilities that emerged from the growing Industrialization, Durand was one of the first architects to clearly define function, in terms of human value, technical performance, and economy, as the key aspect of architectural design. He argues that since building is the most expensive of all arts, yet the sole that is in constant use and thus provides the greatest benefits, it should, in a very Vitruvian manner, be durable, neat, and comfortable, and, in order to keep costs down, symmetrical, regular, and simple (Durand 1831a, b, pp. v, 3). According to Durand, materials appropriate for building are split into three categories: Materials that are hard, difficult to process, and thus expensive, suitable only for buildings of great importance whose public status justifies their usage. Materials that are softer and easier to treat, good for private houses with a smaller budget. And materials whose main purpose is to connect the ones from the previous classes, like iron, copper, lead, but also gypsum, cement, or sand.

Durand distances himself from both the classical assumption that order and proportion are derived from the human body, and the belief in a primitive hut as the archetype for building. As such he neglects the imitation of nature as a design guide and reasons that decoration and ornamentation are temporal matters of taste and, since utterly expensive and not increasing the functionality of a structure, unnecessary and useless. He believes that architecture can be described by certain fundamental principles that are found in any building, regardless of its epoch or style. These elements, once identified, can then be reassembled according to generic methods to create new architecture independent from stylistic considerations, efficient both economically and in terms of design and functionality. By graphically portraying various building types and abstracting their form into simple geometric shapes, he develops a codified scientific method and explicit didactic system that is intended to be applicable without knowing or considering the works of the past (Fig. 2.9).

2.3.1 Durand's Grammars in Relation to Schumacher's Parametricism

Durand's objective of creating a systematic approach for architectural design bears striking similarities to contemporary advocates of computational methods. Patrik Schumacher, Director at Zaha Hadid Architects, describes Durand as "the first to introduce a diagrammatic process within architecture." Durand's simple system of combining basic, standardized elements like walls and columns according to certain rules of alignment, regularity, and symmetry lead to an unprecedented variety of possible results, from which "the rest of the design (including all the familiar classical detail) followed automatically." Schumacher compares Durand's programmatic approach with the compositional processes emerging during Modernism, which became "the true moment of the ordinary diagram." Reasoning further, as an evolution from the ordinary, metric diagram he proposes the extraordinary, parametric diagram, which according to him first appears during the mid-nineties in the use of animation software. Schumacher asserts that parametric modeling allows "to link any parameter/property of any object with arbitrary parameters/properties of any or all other objects within the model." To him the possibilities that arise from setting up dynamic, infinite chains of interdependencies go beyond architectural relevance but are useful "for the organization and articulation of social complexity" (Schumacher 2011, pp. 349, 354).

Concluding from this assumption and reassured in the observation that parametric principles are becoming omnipresent in contemporary architectural design, he proposes a new label for the current period: Parametricism, which according to him could become the first epochal style after Modernism.

> The contemporary style - Parametricism - takes off with the concept that every form in architecture is susceptible to the formulation in terms of continuously varying parameters. [...] Each urban or architectural system is continuously differentiated and functions are scripted that correlate the different differentiations. In this way, deep resonances are established within the overall composition, and the sense of overall organic integration is intensified (Schumacher 2011, pp. 295–297).

The architecture critic Peter Buchanan admits that Parametricism might produce novel and temporarily interesting forms but diminishes its importance to become a successor to Modernism:

> The style can neither adequately frame nor address public space, with facades whose composition and elements allow us to identify and relate to them. Nor do parametric buildings relate to each other (beyond establishing superficial formal contiguities), nor to other architecture (Buchanan 2011).

To Buchanan the potential of parametric modeling is less than the generation of sculptural form but to "bring a wide range of increased efficiencies, in terms of structure, energy, constructional assembly, shaping of flows of people, air, and so on." Buchanan describes Parametricism as a form of 'hypermodernity,' pushing the limits of modernity but not overcoming them. Instead he demands a 'transmodern theory,' which should address global challenges while utilizing a shift toward a sustainable society.

Whether or not the value of both Durand's method of composition as well as current approaches in parametric design is in their ability to create new formal explorations or rather in their use for functional applications, it is interesting to observe that we are currently experiencing a similar frustration with the status quo as did the thinkers of the nineteenth century.

2.3.2 The Intellectual Emancipation of Form from Materiality

Just as Durand, Gottfried Semper shares the opinion that their era is lacking architectural style and like many of his contemporaries he is interested in discovering fundamental architectural principles to formulate new ideas. Yet Semper criticizes Durand's schematic pragmatism and also does not believe that an answer can be found by demanding a return to the origins of architecture, especially not in a Vitruvian sense. To Semper the similarities between the Greek temple, as the paragon of style, and early wood constructions are purely formal and not enough to claim a certain genealogy. In return he develops his own theory on the origins, published in his 1850 book *Die Vier Elemente der Baukunst* (The Four Elements of Architecture), which describes four basic human motifs or necessities. To him the first and 'moral' element of architecture is the fireplace, the hearth, which comes from a necessity for warmth and the heating up of nutrition. Around the central fire the other three elements are formed, the roof, the enclosure, and the terrace, in order to protect it from the remaining and hostile natural elements, wind, water, and earth. Semper claims that all technical proficiencies of humanity can be derived and they originate from these basic elements. Thus the hearth lead to ceramic and metallic manufactures, carpentry origins from the roof, masonry from the terrace, and weaving and textile work through the enclosure (Semper 1851, pp. 55–56). Semper's elements are hence not materially or formally graspable but must be understood as desires or ideas forming from basic needs out of which architecture emerges, in contrast to the Vitruvian notion of imitation. While the elements might be found in nature, the architect creates artificial works; hence he does not imitate but rather translates. In the prolegomena to his in 1878 published work *Der Stil in den Technischen und Tektonischen Künsten* (Style in the Technical and Tectonic Arts, Or, Practical Aesthetics), he explains his theory of style as searching for the components of form, which are not formal in themselves but rather idea, force, means, and matter (Semper 1878, p. viii). From these components he believes that form and architecture will be able to create meaning, transferring purpose through one material to the other, a transformation that he calls 'Stoffwechsel.' Semper, who is quite aware of the scientific progress of his time in other disciplines, borrows the term from biology, where it means metabolism, describing the life-sustaining chemical transformations and exchange of matter within the cells of living organisms. But it can also be understood as 'phase change,' referring to the transitions between solid, liquid, and gaseous states of matter. What Semper means to emphasize however is that primordial human requirements will always remain the same and so will the functionality and purpose of their physical manifestations, even though their form and materiality evolves and changes over time. This emancipation of form from a particular material allows Semper to highlight that the symbolic function of architecture, transferring purpose through one material to another, is as imperative as its material structure since it creates stability on a cultural level (Muecke 2005, p. 30).

Semper's theory marks a decisive moment in architectural thinking. In the past architectural form had often been justified by tracing its evolution back to primitive archetypes that were based on natural principles, representing proportion and order. Similarly, materials were described as having inherent properties, which if discovered determine their correct usage. Others however argued that architecture is largely a formal practice, beauty and proportion a matter of individual taste, and that materials can be categorized and simply applied to fulfill particular demands. Semper does not take any of these positions. He asserts that materiality and spirituality are closely linked, but does not value one over the other since to him the creative idea is at the core of architectural design. Form is thus not a fixed entity but rather a continuously changing and becoming configuration (Semper 1878, pp. viii, xv).

2.3.3 Technological and Material Advancement During the Industrial Revolution

While Semper's *Stoffwechseltheorie* resulted in a certain dematerialization in architectural thinking, untying form, function, and materiality, actual progress in construction and the development of new fabrication and material processes lead to a dematerialization of architecture in its physicality.

In 1851 the Great Exhibition took place in London's Hyde Park, showcasing the latest technological developments of the Industrial Revolution. The structure to house the exposition was the Crystal Palace (Fig. 2.10), a groundbreaking, modular, glass and iron building, designed by the English gardener, Joseph Paxton. The Crystal Palace can be regarded as the first physical outcome of Durand's systematical thinking, even though he did not see iron as an architectural material but solely to connect or decorate others. The glazed structure, which was 564 m long, 124 m wide, and 39 m high, was completely build from standardized parts, which enabled an extremely efficient erection within a very short span of time. But despite the structure being revolutionary in its promotion of homogeneous elements for ease of construction, and its application of new industrial materials like sheet glass in combination with wrought and cast iron, its direct impact on architecture and architects of the time was comparably little.

Even Gottfried Semper, who just after the exhibition ended, wrote a short but essential review on its potential effects for architecture remained fairly reserved. To him the structure was little more than a very versatile "glass-covered vacuum that suits everything one would like to bring into it" (Semper 1852, p. 71). While praising its importance for the technical advancement of glazed open spaces he did not treat it as a work of real architecture and, a bit sarcastically, argued that, since the qualities of iron are its thinness, its ideal would be an 'invisible architecture' (Semper 1878, p. 251).

Fig. 2.10 Joseph Paxton's Crystal Palace in London was entirely covered in transparent glass plates (Queen Victoria opening the 1851 Universal Exhibition, at the Crystal Palace in London, Thomas Abel Prior 1851)

Only a few years after the Great Exhibition, in 1855, Henry Bessemer patented a process to purify liquid pig iron, enabling the mass production of steel, and by this solved various technical issues that so far weakened the material and prevented it from becoming architecturally relevant.

Based on the development of Portland cement in 1824, whose thermal expansion properties are almost identical to those of iron and steel, reinforced concrete emerged in 1867, when the French gardener Joseph Monier experimented with adding iron mesh into concrete flowerpots. Monier quickly improved his artisan process and used it to create water tanks, bridges, and even beams and columns. During the 1870s Thaddeus Hyatt performed numerous empirical tests on reinforced concrete beams and developed the basic groundwork for its further usage (Ballard Bell and Rand 2006, p. 53). However, it was not until 1887 that Mathias Koenen demonstrated how to calculate the necessary reinforcement and provide a scientific basis for the placing of the steel bars (Forty 2012, p. 18).

The superior strength of steel essentially enabled the steel frame structure toward the end of the nineteenth century. Together with the introduction of another technological breakthrough, the Otis safety elevator, the first high-rise buildings appeared. The ten-story

Home Insurance Building in Chicago, build in 1885, was the first tall building that utilized the basic principles of skeleton steel frame construction and its load-bearing opportunities (Leslie 2013, p. 46). In 1931, the 102-story-high Empire State Building in New York was finished as a steel frame structure and remained the tallest building for nearly 40 years. Almost as impressive as the immense height of the building was the speed and execution of its construction, which Rem Koolhaas describes as "a form of automatic architecture". Rising almost a story per day, the building seemed "to generate itself, feeding on the never-ending stream of materials that arrive with split-second regularity" (Koolhaas 1994, pp. 139–141).

2.3.4 Functionality Takes Precedence Over Aesthetics and Form

Notwithstanding the mechanic automation of processes and timely execution of tasks, Le Corbusier, who visited New York for the first time in 1935 was not very affected, by neither the Empire State Building nor the American skyscraper in general, which he provocatively deemed as being 'too small.' To him the American skyscraper was purely decorative, "a plume rising from the face of the city." Instead he proposed the so-called Cartesian skyscraper, a modern and rationally designed tower based on a biaxial cruciform plan.

> The Cartesian skyscraper is a miracle in the urbanization of the cities of machine civilization. It makes possible extraordinary concentrations […] while taking up only 8 to 12 per cent of the ground, 92 to 88 per cent being restored, usable, available for the circulation of pedestrians and cars! These immense free areas, this whole ward in the business section, will become a park. The glass skyscrapers will rise up like crystals, clean and transparent in the midst of the foliage of the trees (Corbusier 1964, p. 83).

Representing a "function of capacity" the monolithic building should be made from a steel skeleton structure, without walls and enclosed in a facade of glass (Corbusier 1964, pp. 53, 81). Laying these towers in a dense but generous grid he made radical proposals for vast urban redevelopments, such as his Ville Voisin, a concept city for Paris with low residential blocks divided by large areas of greenery and parks. Both in its function and form, Le Corbusier's Cartesian skyscraper stands exemplary for his thinking of the house as "a machine for living in" (Corbusier 1986, p. 95).

This blunt but crucial analogy is intended to highlight the necessity for more efficient, modern, and human-focused environments, and suggests adapting processes from engineering and industrial design to overcome formal ideals. By this architects, he believes, will be able to create novel forms that respond to specific needs, without having to claim a relation to earlier architectural styles, since to Le Corbusier style is unnecessary, and although sometimes pleasing, solely decorative (Corbusier 1986, p. 25).

The core of Le Corbusier's ideals are summed up in his manifesto *Five Points Towards a New Architecture* and can be found in many of his buildings (Fig. 2.11). First, the structure should be lifted from the ground, resting upon slender reinforced concrete columns, "the supports." Second, the building should contain a "roof garden," which protects the concrete roof against changing temperatures, creates a luxury outdoor space, and makes up for the green area consumed by the structure. The support system, providing the structural foundation of the building, allows the "free design of the ground plan," meaning that the space can be configured into rooms without the need for supporting walls. The fourth point concerns the "horizontal window," which provides unrestricted views of the surrounding and lights the rooms more efficiently than with vertical windows of the same surface area. The fifth point addresses the "free design of the facade," independent from the location or arrangement of interior rooms. Le Corbusier's five points meant a radical departure from the past and can as such be considered a direct and critical response to Vitruvius' *Ten Books on Architecture* (Corbusier 1970, pp. 99–101).

Fig. 2.11 The roof level of Le Corbusier's Cité Radieuse in Marseille (1947–1952), with the children's paddling pool, atelier, and ventilation stack, exhibits a number of his five points toward a new architecture (Kretzer 2014)

2.4 Synthetic Materials

> The business of Architecture is to establish emotional relationships by means of raw materials.
> Architecture goes beyond utilitarian needs.
> Architecture is a plastic thing.
> The spirit of order, a unity of intention.
> The sense of relationships; architecture deals with quantities.
> Passion can create drama out of inert stone.
> Corbusier (1986, Towards a New Architecture, p. 151)

Le Corbusier's use of the term plastic in the above quote is meant to highlight crucial poetic and sensual qualities of architecture as opposing to solely focusing on utility and construction. Instead of trying to find inspiration in classical and natural forms, he suggests looking forward, to learn from adjacent disciplines and draw from the qualities of machines (Banham 1980, p. 223). Condemning architecture as the only profession "in which progress is not considered necessary" he promotes the use of composite and synthetic compounds and demands an architecture that reaches beyond purely utilitarian needs but instead embraces technological evolution and new materials (Corbusier, pp. 109, 215, 229).

The first entirely artificial material Bakelite plastic was invented in 1907 by Leo Baekeland. Its astounding properties quickly lead to its intense use by the United States military as an alternative to steel for the production of lightweight weaponry. During the 1920s, in the post WWI era, the North American economy expanded massively and boosted further industrial research which lead to the invention of new types of plastics at a regular basis, including Nylon, Teflon, PVC, or Vinyl. In 1933 Eric Fawcett and Reginald Gibson accidentally discovered polyethylene (PE), which continues to be the most produced and highest selling plastic in the world today (Ballard Bell and Rand, p. 220).

Richard Buckminster Fuller was one of the first to propose plastics in his mobile and eco-friendly scenarios, like the Dymaxion House (1927), the Wichita House (1945), or the iconic Montreal Biosphere for the Expo 1967 (Fig. 2.12).

> In respect to physical resources, until recently man had assumed that he could produce his buildings, machinery, and other products only out of the known materials. […] But now in the aerospace technology man has developed his metaphysical capabilities to so advanced a degree that he is evolving utterly unique materials "on order." Those new materials satisfy the prespecified physical behavior characteristics which transcend those of any substance previously known to exist anywhere in the universe. Here we witness mind over matter and humanity's escape from the limitations of his exclusive identity only with some sovereignized circumscribed geographical locality (Buckminster-Fuller 1969, p. 32).

Fig. 2.12 Buckminster Fuller's Montreal Biosphere was originally enclosed with acrylic cells and incorporated a dynamic shading system to control its internal temperature (Kretzer 2009)

Another remarkable pioneer during these times was the Austrian-American architect, Frederick Kiesler, who already in 1925 formulated a series of architectural demands as an opposition to the mostly functionalist architecture of his time. He demanded:

1. Transformation of the surrounding area of space into cities.
2. Liberation from the ground, abolition of the static axis.
3. No walls, no foundations.
4. A system of spans (tension) in free space.
5. Creation of new kinds of living, and, through them, the demands which will remould society (Kiesler 1925, pp. 141ff).

Kiesler's five points became the theoretical foundation for his Endless House project, a speculative study that aimed to create a spatial symbiosis between man, nature, and technology (Bogner and Noever 2001, p. 11). To him space and time were continuous and elastic and should thus be interwoven and mutually dependent. Moreover, architecture should be capable to provide ideal solutions to the varying social demands and uses of its occupants. As a physical manifestation of his ideas he designed the Space House in 1933 for the New Yorker Company Modernage Furniture, which was supposed to be cast entirely in plastics to create a fluid transition between floors, walls, columns, and ceilings.

Kiesler believed that his ideas were very influential on architects of his time, like Le Corbusier or Mies van der Rohe, yet his projects were received rather dismissively (Borden and Rendell 2000, p. 62). This however only strengthened his desire to break out of the ordinariness, especially the International Style, and propose ever more radical and futuristic visions. But even though he continued to develop his ideas in constant response to technological and social trends, the Space House remains the only building he ever realized, a fact which in 1960 lead Philip Johnson to call him "the greatest nonbuilding architect of our time" (Johnson 1960, p. 70).

2.4.1 The Age of Plastics

After the Second World War, due to a lack of conventional building materials and a shortage in housing, research into petroleum chemistry seriously caught fire, and various types of plastic were brought into commercial application. In the final chapter of their book *Plastics*, published in 1941, the British chemists Victor Yarsley and Edward Couzens, vividly anticipate an Utopian vision of the 'Plastic Age' that was to follow:

> This 'Plastic Man' will come into a world of colour and bright shining surfaces, where childish hands find nothing to break, no sharp edges or corners to cut or graze, no crevices to harbour dirt or germs, because, being child, his parents will see to it that he is surrounded on every side by this tough, safe, clean material which human thought has created. The walls of his nursery, all the articles of his bath and certain other necessitates of his small life, all his toys, his cot [...] all will be plastic. [...] As he grows he cleans his teeth and brushes his hair with plastic brushes, clothes himself with in plastic clothes [...] The windows of his school [...], like those of his house are of moulded plastic, light and easy to open never requiring any paint (Yarsley and Couzens 1941, p. 149).

The first practical architectural experiments focused on the development of prefabricated elements to create mass-produced shelter. But it was not until the early 1950s, when a new generation of plastics appeared, that architects started to really develop ideas that did justice to the materials' various properties. At a 1954 conference on the use of plastics in building Robert K. Mueller from the Monsanto Chemical Company compares the impact of new technologies and materials on the reduction of costs in automobile construction and predicts dramatic advances in building methods and materials:

> The trend towards prefabricated building structures and structural elements presents an unusual opportunity for plastic materials. Plastics are capable of revising the "architectural index" of our time. We predict that plastics engineering will play a significant role in a new American style of building architecture because of inherent features of plastic materials and their adaptability in any type of design (Mueller 1955, p. 127).

The visionary thinking of the company, back then a major global producer of plastics, was manifested in a collaborative project with Disney Imagineering and the MIT from 1953 to 1956, the Monsanto 'House of the Future.' The almost

5-m wide building, with its four square rooms cantilevering outward from a central core, was made entirely from glass fiber-reinforced plastic (GFRP) shells and showcased a vision of the year 1986. The center of the structure housed the kitchen and two bathrooms, while the living room, family room, master bedroom, and two small bedrooms occupied the wings. According to a press release from the Monsanto Chemical Company in 1960, the influence of the experiment on the building industry was remarkable:

> All told, 23 per cent of the plastics made in this country now go into construction, compared with 15 per cent the year before Monsanto's experimental house was built. The pace being what it is, the "Plastics Home of the Future" may soon be just one of many plastics homes of the present (Weiss 2010).

Although Monsanto's vision of plastic homes did not quite come true as they imagined, the usage of polymers in building construction has continuously risen, making it the second largest consumer of plastics today, just after packaging (Fernandez 2006, p. 161). Ironically, it was precisely the house's futuristic and style-dependent use of technologies and materials that lead to its abolition only 10 years after its opening. While the high-tech interior simply became standard and out of fashion, the use of fiber-reinforced polyester for the shell structure failed as a building material because it conversely was too durable and resistant for a time that valued temporality and progress.

2.4.2 Utopian Visions for a Better Tomorrow

In 1958, motivated by the results of the 1956 CIAM congress held in Dubrovnic, which addressed issues of mobility, flexibility, growth, and communication, a group of young architects around Yona Friedman, founded the 'Groupe d'Études d'Architecture Mobile (GEAM).' Centered on the understanding that people should be actively engaged in the process of constant sociological and technological change, the group proposed the so-called 'architecture mobile,' which was to consist of flexible room modules without specific functions that could be dynamically inserted into load-bearing infrastructures (Jeska 2008, p. 12). Friedman explains:

> The two techniques I proposed were aimed to liberate the building from the groundwork. The elements were the 'span-over blocks,' supported by stair towers which were about 60 meters away from each other and the 'raft blocks,' boxes placed on the ground which rested on 'beams.' [...] I express this philosophy accepting the unpredictability of human behavior, of the inhabitants, also accepting the illusion of planning, the erratic character of each person's story. I assume that in every domain, the process is important and a final result which can be determined with absolute certainty does not exist (Friedman 2006, pp. 14–15).

The formation of the group and its visionary and radical philosophy were characteristic for the period of the early 1960s. Driven by a euphoric decision to abandon traditional practice, a fascination for space travel, and a strong belief in technological progress and the usage of uncommon materials, numerous artistic collectives like Archigram (1960), E.A.T. (1967), Haus Rucker Co. (1967), Coop Himmelb(l)au (1968), or AntFarm (1968) formed around the globe and developed architectural concepts based on mobility, individualism, and self-expression. For them architecture had to go beyond the creation of isolated structures and tend toward environmentally and organically funded design (Scott 2008, p. 62). They relied on a future of abundant resources and developed visionary urban scenarios and large spatial experiments that tried to escape from the limits of established norms but wanted to generate a more flexible and independent living (Sadler 2005, p. 95).

Plastic turned into a term that stood for far more than simply a material. It became eponymous for a complete era, a time and people that were elastic and dynamic but also slick and clean, just as Andy Warhol famously noted:

> I love Los Angeles, and I love Hollywood. They're beautiful. Everybody's plastic, but I love plastic. I want to be plastic (Lavin 2014, p. 15).

The World Expo 1970 in Osaka with its theme 'Progress and Harmony for Mankind' marks to date the climax of plastic architectural exploration and experimentation. Showcasing pneumatic structures, media pavilions, and stretched skin systems it represented a cheerful and optimistic vision of the material's outstanding properties and possible structural manifestations. Despite the public success of these case studies and experiments, the first real plastic buildings on the market however turned out to be the exact opposite, creating anonymous, inhuman, and impersonal environments, with no room for individuality and self-expression. In junction with the oil crisis in 1973 this sudden realization meant the abrupt end of the plastics euphoria demanding a return to natural and more human materials (Jeska 2008, p. 22).

2.5 Intermediate Summary and Conclusion

Architecture is generally a rather slow and conservative discipline. New material innovations thus often take much longer until they lead to substantial changes and advancements. Steel, for example, was in the beginning only used in works of engineering, such as bridges or towers, before it turned architecturally relevant due to its amalgamation with concrete. Plastics on the other hand were treated as a highly imperative structural material but in the end had a much more subliminal influence and are today mostly found in technical appliances and infrastructures. The impact of a material on building is thus not only related to technological advancements but equally to various associative qualities, differing according to culture, society, fashion, and time (Ashby et al. 2009, p. 46). Obviously, none of the above-described periods can simply be reduced to the small number of positions that are portrayed, neither can these opinions be fully grasped in the brevity of the present discourse, yet there appear to be underlying, recurring patterns driving, and strengthening the various stages.

Architects of the Enlightenment, such as Perrault, Laugier, or Ledoux, for example, relate strongly to thinking during the Renaissance, in particular the writings of Alberti, while expressively demarcating themselves from the previous Baroque and Rococo era. The Renaissance period in turn is clearly defined by a rediscovery of the work and philosophy of the Roman time, as is the Romanesque, and tries to express its difference to the earlier Gothic buildings. At the same time the intricate and decorative architecture of the Gothic era can be associated with the ornamental Baroque, Rococo, and the later Romanticism. When displaying this observation graphically it can be depicted as an ascending curve Fig. 2.13, divided into two distinctive sides. The left part of the diagram displays the styles that aim for structural forms and functional spaces, the right side is more playful in the addition of ornament and decoration. As styles progress they often try to demarcate themselves from the directly previous period while relating to the one before. The graph also displays that the time span of how long it takes for the respective styles to be superseded by a more modern approach gets continuously shorter. While Roman architecture lasted for almost 650 years, the Gothic style was deprecated after 350 years, Neoclassicism after 200 years, and Modernism already after 60 years.

Lbeit this is clearly a hypothetical assumption, it can be exemplified further through one particular material: concrete. During Roman times concrete was used pragmatically as a structural material, but remained mostly hidden behind more appreciated materials, such as brick or marble. While concrete disappeared after 500 AD, sandstone took its place, which during the Gothic era was treated sculpturally and during the Renaissance rather reductive and restrained. The use of cement slowly returned in the mid seventeenth century, this time largely in the form of decorative stucco during the Baroque. With Industrialization reinforced concrete emerged and again its structural qualities prevailed. Le Corbusier emphasized the plasticity and malleability of concrete, yet used it in a very clear and formal manner. The same applies for the spatial explorations in plastics during the 1950s and 1960s.

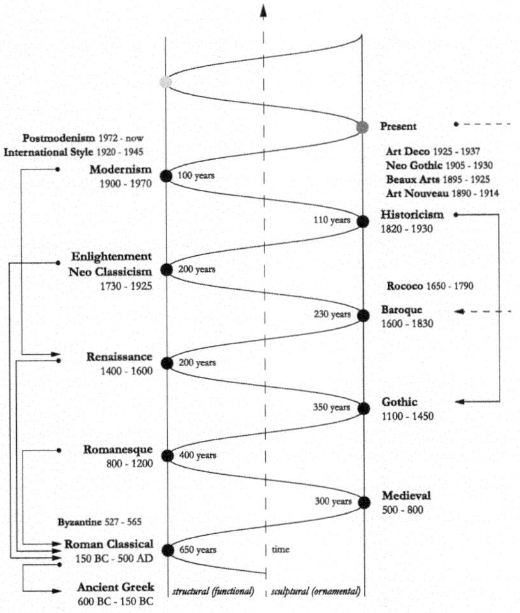

Fig. 2.13 Graphical depiction of alternating architectural styles in form of an ascending curve with arrows showing various relations (Kretzer 2015)

Very recently, we are experiencing a new sculptural approach toward the same material, enabled by novel manufacturing techniques and more sophisticated and powerful computer hardware and software. One very prominent example is the study Digital Grotesque (Fig. 2.14) by Michael Hansmeyer and Benjamin Dillenburger at the Chair for CAAD, ETH Zurich, where complex algorithmic processes in combination with cutting edge 3D sandstone printing are applied to create intricate, uncanny computational forms and geometries.

Fig. 2.14 The 3D printed space Digital Grotesque explores the current limits of computational design- and fabrication technologies (Hansmeyer and Dillenburger 2014)

2.6 Digital Materials

> Digital machines and productive technologies in general allow for the production of an industrial continuum. From the mold we move toward modulation. We no longer apply a preset form on inert matter, but layout the parameters of a surface of variable curvature. A milling machine that is commanded numerically does not regulate itself according to the build of the machine; it rather describes the variable curvature of a surface of possibility. The image-machine organization is reversed: the design of the object is no longer subordinated to mechanical geometry; it is the machine that is directly integrated into the technology of a synthesized image.
>
> Cache (1995 Earth Moves: the Furnishing of Territories, pp. 96–97)

As described in the previous chapters, every new technological development has not only affected the creation and design of architecture but equally architectural thinking and theory. Around the turn of the twenty-first century the growing prominence of computer-aided design tools paired with influences from computer graphics and especially motion and cinematic animation lead to a number of critical investigations into their greater relevance in relation to future architectural explorations. In the *AD—Architectural Design* issue *Folding in Architecture*, edited by Greg Lynn and first published in 1993, Lynn argues that architects during the preceding two decades had been largely focusing on the creation of "heterogeneous, fragmented and conflicting formal systems" resulting "from a logic which tends to identify the potential contradictions between dissimilar elements." Emerging as a reactionary response to this formal dispute he addresses a second tendency that aims at recovering unified architectural languages either through historical reference (Neoclassicism or Neomodernism) or by finding local consistencies (Regionalism). Yet Lynn contends:

> Neither the reactionary call for unity nor the avant-garde dismantling of it through the identification of internal contradictions seems adequate as

> a model for contemporary architecture and urbanism (Lynn 1993, p. 24).

As a more appropriate alternative he proposes a post-contradictory work based on smooth transformations that allow the incorporation of variety within an incessant but heterogeneous whole. Referring to Deleuze's understanding of smoothness as "continuous development of form" Lynn reasons that through smoothness and pliancy architecture can address complexity through flexibility, which will allow the integration of "unrelated elements within a new continuous mixture." Mentioning both the Gehry House (1991) and Peter Eisenman's Wexner Center (1989) he implies that despite their clear deconstructivist appearance they express a certain softening toward curvilinear deformations that foster "a more fluid logic of connectivity" rather than emphasizing contradiction and conflict. From this observation he argues for the development of an architecture that bends and folds locations, materials, and programs rather than breaking and disrupting them, to create fluid and dynamic systems that maintain both the individual characteristics of each element and remain open for future additions (Lynn 1993, pp. 24–28).

In Bernard Cache's *Earth Moves*, Michael Speaks builds upon Lynn's argumentation but wonders whether a shift from deconstructivist forms toward folded forms is enough in order to be called new and important or if it does "simply repeat what already exists." Contending that Deleuze's work is much more comprehensive than what Lynn emphasizes, Speaks bemoans that by simply picking certain theoretical or philosophical ideas and translating them into architectural shapes, "architecture becomes applied philosophy, and necessarily gives up all claims to singularity and creativity." To Speaks the real value of the Deleuzian concept of the fold lies in the "shaping of the form of practices (including techniques and logics), rather than the shaping of individual architectural forms." In that sense he demands a more abstract understanding and translation of the Deleuzian notion:

> Thus new forms of architecture will not emerge as a result of the effects achieved by ever more pliant, fluid, complex, and heterogeneous shapes or architectural forms, but rather with the development of more pliant, complex, and heterogeneous forms of architectural practice-with architectural practices supple enough to be formed by what is outside or external to them, yet resilient enough to retain their coherence as architecture (Speaks 1995, pp. xv, xvi).

Bernard Cache has according to Speaks succeeded in developing a conversion from theory toward practice, by employing the fold as a method to reconsider the relationship of the interior with the exterior world through images. Cache has early on been strongly engaged in what he calls Computer-Assisted Conception and Fabrication (CFAO) systems. He identifies two main types of CFAO usage in the industry, mainly in the areas of mechanical engineering and automobile or aeronautical applications, which albeit they have "increased the productivity of the idea" they so far do not offer any fundamental "advances over the work done by hand." Growing from this he imagines second-generation systems where objects are not any more designed but calculated, allowing the creation of complex forms that would be difficult or impossible to draw in a traditional manner. More interestingly however he asserts that such systems might incept a "nonstandard mode of production" where objects from the same series can vary in size and shape by simply modifying certain fabrication parameters on the fly. Since neither the function of an object nor its materiality are any longer linked to a particular form but can vary continuously, the digital representation "takes precedence over the object." The primary depiction of an object is hence not anymore its image but a model of simulation, of numerical manipulation, resulting in a kind of digital materiality (Cache 1995, pp. 88–98).

> The shape of this new objectivity prolongs surfaces of resonance, whether screens or membranes, that restore the materiality of the numerical processes. Data of this sort can then create an image on a cathodic screen, but it can also create a sound on an acoustic membrane or, better still, produce a surface of variable curvature (Cache 1995, p. 97).

Fabio Gramazio and Matthias Kohler at ETH Zurich build upon this notion and use the term digital materiality to describe materials that are increasingly being enhanced with information technological characteristics:

> Digital materiality evolves through the interplay between digital and material processes in design and construction. The synthesis of two seemingly distinct worlds - the digital and the material - generates new, self-evident realities. Data and material, programming and construction are interwoven. This synthesis is enabled by the techniques of digital fabrication, which allows the architect to control the manufacturing process through design data. Material is thus enriched by information; material becomes 'informed' (Gramazio and Kohler 2008, p. 7).

In contrast to materials that only exist in digital or virtual form, such as textures for 3D renderings or computer software that simulates material properties for evaluation or visualization, the present meaning of *physical* digital materials is thus directly tied to the emergence and evolution of digital design and fabrication techniques.

In *Beyond the Grid*, Ludger Hovestadt compares the development of architectural manufacture to the history of biology: While during the seventeenth century nature was described in rather broad terms and with the help of patterns, the invention of the microscope allowed the study of individual cells and thus not only facilitated the decomposition of biological forms but also their schematic reconstruction. A similar leap happened in the architectural world during the eighteenth century when "production became a necessary condition for architecture" in contrast to the slow and laborious building techniques of the previous times. A little later, during Industrialization, the standardization of building elements paired with new technological processes made possible unprecedented architectural explorations as exemplified in the previous chapters.

In biology another crucial breakthrough occurred with the discovery of DNA in the 1950s. Since then the internal code of cells can be revealed and in a very moderate way be recombined and reassembled. Around almost the exact same time lie the roots of computation and digital fabrication (Hovestadt 2009, p. 19).

2.6.1 The Emergence of Digital Design and Fabrication Techniques

In 1949 the United States Air Force assigned John Parsons to develop an economical, automated method for the manufacture of helicopter rotor blades. Parsons teamed up with the Servomechanisms Laboratory at MIT, who back then was on the global forefront of mechanical computing and feedback systems, and together they created the computer-controlled 'Card-a-Matic Milling Machine' for three-axis contour milling. Although the machine required relatively long in preparation and could only perform a limited set of operations that were fed from punch card tapes, it represented a revolution in automated manufacturing. The machine reduced workforce, downtime, and waste material while increasing productivity, precision, and especially versatility through the use of a single tool. However, the industry was hard to convince of the machine's potential and remained unwilling to provide the necessary funding for further industrial development (Caneparo 2013, p. 55).

During the 1960s computing became much more publicly prominent and a new generation of cyberneticists began actively speculating about architectural design and its similarities to cybernetic systems (Spiller 2008, p. 10). In 1963 Ivan Sutherland released Sketchpad (Fig. 2.15), a pioneering attempt for human–computer interaction, often credited as the very first computer-aided design (CAD) tool. Sutherland himself describes Sketchpad as a system that, "by eliminating statements (except for legends) in favor of line drawings, opens up a new area of man–machine communication," which in the past "has been slowed down by the need to reduce all communication to written statements that can be typed." The tool ran on a Lincoln TX-2 computer, a highly advanced machine at the time, and incorporated a light pen that was used to directly draw lines and other geometrical shapes onto a

Fig. 2.15 Ivan Sutherland using Sketchpad with a light pen to directly draw geometrical shapes onto the computer (scanned by Kerry Rodden from original photograph by Ivan Sutherland 1963)

small connected cathode ray tube. Additional knobs and buttons allowed the further manipulation of the content in real time and an adjacent plotter was used to print the created diagrams.

Sutherland saw the advantages of his invention especially in the creation of drawings where motion, analysis, high accuracy, or repetition were essential, yet he conceded that "it is only worthwhile to make drawings on the computer if you get something more out of the drawing than just a drawing" (Sutherland 2003, pp. 17, 99, 110).

Graphical CAD systems were further improved in the 1970s and became suitable tools for designers and architects for the creation of geometric representations. Simultaneously, numerical control (NC) systems were enhanced to add visual descriptions of the work piece and process. The new software was called computer-aided manufacturing (CAM) and the immediate connection to CAD helped that it was quickly adopted by various industries for the creation of complex products like ships or automobiles. While in the beginning the technology remained only viable for projects with large turnovers and as such big companies, continuous increase in computing power and the concomitant reduction in costs eventually lead to the propagation of CAD/CAM technologies and CNC fabrication in the areas of industrial design and manufacturing (Corser 2010, p. 13). In parallel to the development of subtractive fabrication processes such as milling or cutting, additive manufacturing, today best known as 3D printing, came on the market in the 1980s, and was rapidly promoted by the emergence of numerous companies that pushed its development (Gershenfeld 2012, p. 45). Around the same time industrial robots became more prominent and sophisticated in the range of movements and actions they could perform. Due to their great versatility since the "end effectors, attached and controlled by these arms, are as diverse as the materials they can process" much of today's research into digital fabrication focuses on the potential of robotic arms (Beorkrem 2013, p. 10).

Neil Gershenfeld, head of MIT's Center for Bits and Atoms, argues that the cumulative emergence and evolution of these fabrication technologies will inevitably result in a 'new digital revolution,' empowering people "to design and produce tangible objects on demand, wherever and whenever they need them." Drawing analogies to the history of computing he predicts that through the continuous increase in performance and versatility of, for example, desktop 3D printers and the open access to their blueprints, tool owners can not only use but also reproduce and alter them, just as the early personal computer allowed people to create their own software. Encouraged by the success of **Fab Labs** he believes that the power of these technologies lies in their largely unregulated but well connected nature, thus their real strength "is not technical; it is social." Gershenfeld suggests that the ultimate abilities of what he calls 'assemblers,' the future descendants of 3D printers, will be "to create complete functional systems in a single process" without the creation of any waste or trash (Gershenfeld 2012, pp. 43, 52, 57).

2.6.2 New Instruments for the Architectural Design Practice

Despite the obvious advantages of digital technologies for design and manufacture and despite the fact that code already represents the base and framework for any kind of planning, communication, financing, infrastructure, and construction, as Hovestadt observes, the building industry has to date been fairly slow and repellent in adapting (Hovestadt 2009, p. 19). This is due certainly to the fact that it mostly deals with one-off projects and relatively small profit margins, however the reason might also be rooted in an outdated understanding of architectural design and production based on tradition and craftsmanship. Referring to William Mitchel, Branko Kolarevic, Professor at the University of Calgary, states that while in the past "architects drew what they could build, and built what they could draw" we have now reached a state where the constructibility of a design, enabled through digital production processes, becomes directly linked to the limits of computation. Due to this:

> The question is no longer whether a particular form is buildable, but what new instruments of practice are needed to take advantage of the opportunities opened up by the digital models of production (Kolarevic 2003, pp. 31–33).

Fabio Gramazio and Mathias Kohler add that in order "to make the full spectrum of digital technologies in architecture accessible [...] they have to be considered conceptually in design from the very beginning." Hence they argue for research and practice-based approaches that go beyond investigating the potential of technologies but focus on "integrating them in an early design phase in order to finally overcome the still prevalent separation of design and making and introduce new meaning and substance into the profession" (Sheil and Glynn 2011, pp. 2, 6).

An early pioneering application in using computer-aided design and engineering for architectural fabrication can be witnessed in the Sydney Opera House, designed by the Danish architect Jorn Utzon and realized from 1956 to 1973 (Fig. 2.16). Digital modeling was applied for both structural analyses as well as the detailed layout of building assemblies and inventory documentation. Additional computer-generated information was used for factory production and plotted templates were created for on-site cutting of large glass panels. While the majority of the work was still based on manual labor, the project would not have been realizable without the application of computers to solve and simplify the complex geometry. The Sydney Opera House thus marks a building utilizing and progressing technologies way ahead of its time.

Fig. 2.16 Sydney Opera House by the Danish architect Jørn Utzon behind Harbour Bridge, Sydney, realized from 1956 to 1973 (Kretzer 2005)

One of the first architectural projects that was developed and also produced largely digital is the Great Fish Sculpture by Frank Ghery at the entrance to the Vila Olimpica in Barcelona, Spain, built in 1992 (Fig. 2.17). The structure, which is 55 m long and 35 m high, is made of interwoven steel strips that are connected to an open steel framework. Since back then no architectural computer software existed that would allow for both the design and the production of the intricate form, the architects leaned toward the aerospace industry and found a

Fig. 2.17 Frank Ghery's Great Fish Sculpture at the entrance to Villa Olimpica, Barcelona, built in 1992 (Kretzer 2016)

solution in the program CATIA. In what Kolarevic calls "a radical departure from the normative practices of the profession," the tool was used for the design, structural analysis, fabrication, and as on-site construction guide (Kolarevic 2003, p. 31).

The realization that complex surfaces and geometries consisting of multiple unique pieces can be manufactured in a digitally controlled way without significantly increasing the cost of fabrication in comparison to mass production motivated Ghery and Partners to establish Ghery Technologies in 2002. The company since then develops and promotes Digital Project, a version of CATIA, tailored to the specific demands of multifaceted architectural projects including the preparation of fabrication data. The work of Frank Ghery is thus not only revolutionary in its architectural complexity and appearance but according to Lisa Iwamoto, founding partner of Iwamoto Scott in San Francisco, also in expanding "the role of the architect to include oversight of the building and construction-management process, much as it was in the age of the master builder" (Iwamoto 2009, p. 6).

2.6.3 The Digital Chain: From Design to Production

Today, digital design and fabrication technologies have infiltrated almost every architectural school on various levels and many of the larger architectural offices have established their own specialized computation units and in-house fabrication and prototyping facilities, slowly arriving at what Kolarevic calls "a digital continuum from design to production" (Kolarevic 2003, p. 7). Bob Sheil, Director of The Bartlett School of Architecture, consents and describes a paradigm shift arising from the direct engagement with digital technologies, which empower the architect to become involved in areas that he has formerly been excluded from. Sheil reasons that in the past the making of buildings was inherently dependent on the successful communication between the various involved disciplines, often resulting in negotiated translation of the original design intent, depending on the complexity of transfer from one area to the other. However through the integration of digital technologies "the exchange of information between design and fabrication is no longer a slow chain of vulnerable links, but a rapid flow of data, where design and making can be a simultaneous process," redefining the role of architects into 'hybrid disciplinarians' (Sheil and Glynn 2011, p. 156). Hovestadt describes a similar phenomenon when referring to the digital chain:

> The digital chain represents the general formulation of the building process, its reference process, including all of its abstract building blocks. Form and its design become a variable that - freed from the dependency on function - can now focus on their essence, that is, representation (Hovestadt 2009, p. 133).

Yet, while new algorithmically computed forms are continuously emerging and while digitally controlled tools become increasingly accessible, materials still have fairly little design influencing effects and are often shifted to the end of the process where they undergo solidification within production. As a result contemporary architects often think of materials as immaterial components, which can be chosen from a design palette and, similar to the use of textures in 3D renderings, applied as visual and compositional architectural elements. This trend of categorizing and sorting materials into sets of quantifiable qualities becomes evermore obvious in the large amount of material databases, libraries, and catalogs that appear in print, online, and physically in every larger city.

Criticizing the "impoverished notion of form generation, which refers to various digitally driven processes resulting in shapes that remain detached from material and construction logics," Achim Menges and Michael Hensel demand a more holistic understanding of form, material, structure, and behavior (Hensel and Menges 2008, pp. 55–56). According to Mette Ramsgard Thomsen, head of CITA (Centre for Information Technology and Architecture) at the Royal Danish Academy of Fine Arts, this will strengthen the reemergence of 'good building culture' and help to overcome the loss of uniqueness and shift toward the general and mass-produced, which came from the industrialization of construction processes (Ayres et al. 2012, p. 8).

However, there are two things that tend to be neglected. First, for computer-controlled machines to analyze and apprehend the particular materials they are treating they would need to be equipped with some sort of feedback and evaluation systems. Second, they would have to be able to learn and respond to observed material variances by improvising and adjusting their fabrication strategy. Until this is achieved they will always require super-standardized and homogeneous materials, since albeit the executed steps may differ the operational principle remains the same. So while the complexity of possible material handling increases with the versatility of the tool, the homogeneity of the material has to be raised accordingly, leading to a certain dematerialization of material-specific characteristics, especially when dealing with nonsynthetic substances. The decisive properties of wood, for example, like its fiber orientation, texture, varying density, strength, and color are hardly tolerable for automated manufacture and thus have largely been eliminated by the creation of various wood composite materials such as chipboard, medium-density fiberboard (MDF), or oriented strand board (OSB) (Schindler 2009, pp. 173–179, 207). What this all amounts to is that digital fabrication techniques may promote the departure from the *industrialization of construction processes* but at the same time require the increased *homogenization of material properties*.

2.7 Information Materials

> The mechanical brain does not secrete thought "as the liver does bile," as the earlier materialists claimed, nor does it put it out in the form of energy, as the muscle puts out its activity. Information is information, not matter or energy. No materialism which does not admit this can survive at the present day.
>
> Wiener (1961, Cybernetics, p. 132)

The term *information materials* aims at describing a type of newly emerging materials which share certain commonalities that distinguish them from the traditional notion of materiality.

- First, *information materials* have the inherent capability to contain and harvest (digital) information and transform it into physical representation. Thus they are dynamic and can change their state over time in a controlled way and in response to external influences.
- Second, *information materials* are based on information technology. They are artificially created on a symbolic level by the combination of formerly distinct elements into functional assemblies using digital technologies. As such they are not built upon anything that can be found in nature but are sole products of human intellect (Bühlmann and Hovestadt 2013, p. 11).

The new term *information materials*, as a distinction from other names, such as programmable matter or smart materials, which will be discussed in the following, wants to establish a general awareness of these phenomena and develop a linguistic proficiency among the architecture and design community. The concept thus aims at proposing and mediating a new way of thinking, liberated from a materialistic and mechanistic point of view, and instead focusing on *materiability*, the empowering ability to create synthetic materials with performative abilities.

2.7.1 Programmable Matter

Programmable matter generally refers to matter or materials that are capable of changing their physical properties, such as shape, density, conductivity, opacity, etc., in a controlled and direct way. The term was introduced by Tommaso Toffoli and Norman Margolus, describing the concept of a three-dimensional network of fine-grained functional substrates. These minute elements are able to interact with each other and through varying their formation, arrangement, or individual properties, produce a dynamic, new kind of synthetic material. More precisely, Toffoli and Margolus define four essential key aspects of what they understand as programmable matter:

> (i) It can be assembled into lumps of arbitrary size (the limits being given by economics rather than technology).
>
> (ii) It can be dynamically reconfigured into any uniform, polynomially interconnected, fine grained computing network.
>
> (iii) It can be interactively driven, in the sense that its dynamical evolution can be started, interrupted, and resumed at any moment in response to the occurrence of specified internal or external events.
>
> (iv) It is totally accessible to real-time observation, analysis, and modification (Toffoli and Margolus 1991, p. 266).

What they essentially aim for is the creation of individual artificial computing nodes at a scale small enough in order to accumulate and self-assemble into any kind of materiality. Building upon this work, Seth Goldstein and Todd Mowry initiated the Claytronics project at Carnegie Mellon University, researching the production of nanoscale computers and robots, referred to as claytronic atoms, or catoms that can form tangible three-dimensional objects that users can interact with. Each catom is in its preliminary design "a unit with a CPU, a network device, a single-pixel display, one or more sensors, a means of locomotion, and a mechanism for adhering to other catoms" (Goldstein et al. 2005, p. 99).

Once the work succeeds in overcoming some crucial aspects, such as shrinking the elements, which are still in the centimeter range, to millimeter or even micrometer sized particles, providing the individual nodes with sufficient energy while maintaining their lightness and versatility, and developing stable software and hardware configurations that function in ensembles of millions of catoms, Goldstein and Mowry imagine:

> Programmable matter will allow us to take a (big) step beyond virtual reality, to synthetic reality, an environment in which all the objects in a user's environment (including the ones inserted by the computer) are physically realized (Goldstein and Mowry 2004).

The possibilities that such an environment would open up are nearly infinite, and even though the concept still sounds more like something from a science fiction movie, researchers in the domain are convinced that it is only a question of time and persistence until it becomes physical reality (Guin 2012).

While both Goldstein and Mowry as well as Toffoli and Margolus imagine to build up some sort of universal material from a multitude of minute controllable elements another approach toward *information materials* involves the creation of multifunctional materials with specific dynamic properties, also referred to as smart materials. Although these materials certainly lack the versatility of the former approach, they offer various advantages including the departure from mechanics and engineering toward the areas of materials science, chemistry, and biology and as such a much softer domain of research.

2.7.2 The Soft Kinetics of Smart Materials

The director of MIT's Self-Assembly Lab, Skylar Tibbits, investigates in what he calls 4D printing, which "entails multimaterial prints with the capability to transform over time, or a customized material system that can change from one shape to another, directly off the print bed." The material that is being used for the active parts of his transformable assemblies is a hydrophilic polymer, which is able to expand 150 % upon immersion in water. In combination with a rigid polymer that provides the required stability the printed systems can then self-change from one state to another when put in contact with water. While Tibbits' method in 3D printing smart materials undoubtedly offers various exciting opportunities, which he finds mostly "for the future of the products, and the shipping and manufacturing sectors," the creation and application of shape changing smart materials is not new (Tibbits 2014, pp. 119–121). Among the best-known and widest-used materials are so-called bimetals, which due to the differential expansion of two tightly bonded metals are able to flex reversely in response to a change in temperature. This property has among others been extensively used for the actuation of on/off switches within early thermostats and is depicted in the following graphic (Fig. 2.18).

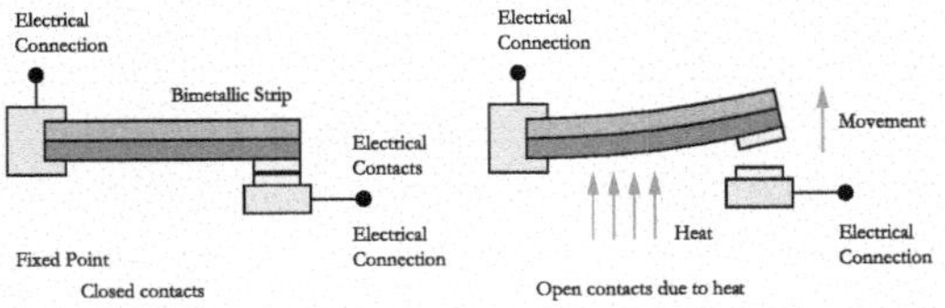

Fig. 2.18 Schematic diagram of the working the diagram appears a little too big principle of a bimetallic strip as used in early thermostats (Kretzer 2015)

The English scientist Gordon Pask was one of the earliest to describe the puzzling effect of the self-activated material in 1972:

> It seems to me that the notion of machine that was current in the course of the Industrial Revolution - and which we might have inherited - is a notion, essentially, of a machine without goal, it had no goal 'of', it had a goal 'for'. And this gradually developed into the notion of machines with goals 'of', like thermostats, which I might begin to object to because they might compete with me. Now we've got the notion of a machine with an underspecified goal, the system that evolves. This is a new notion, nothing like the notion of machines that was current in the Industrial Revolution, absolutely nothing like it. It is, if you like, a much more biological notion, maybe I'm wrong to call such a thing a machine; I gave that label to it because I like to realise things as artifacts, but you might not call the system a machine, you might call it something else (Haque 2007, p. 54).

Pask was a convinced advocate of cybernetics, a term coined by Norbert Wiener in 1948, describing systems that are able to assess feedback on their state and progress and alter their course accordingly. To Pask the thermostat represents a very basic cybernetic system that, when the temperature of a space goes beyond a certain threshold, will automatically either induce a heating or cooling process in order to retain the previously set target temperature. It is interesting to observe Pask's trouble in properly describing and defining the phenomenon of both the cybernetic system as well as the dynamic material. He clearly distinguishes it from the mechanical notion of machines and refers toward more biological and organic associations. By this Pask already anticipated some of the difficulties we are still facing today when dealing and working with *information materials.*

2.7.3 Definitions of Smart Materials

The term smart materials is a relatively young invention and generally refers to materials that can change their properties in response to environmental conditions, such as changes in temperature, humidity, pressure, stress, pH level, magnetic or electrical field, light radiation, and many more. The degree of smartness or behavioral response of a certain material is often measured by comparing the amount of transformation to the duration it takes to complete a full cycle. Smart materials are generally classified according to their basic physical or chemical effects. Table 2.1 shows an overview of several smart materials displaying both the stimulus applied and the actual response of the material.

The history of smart materials differs according to the respective type of material and the first recorded occurrences of certain effects can in cases date back centuries, like, for example, the discovery of Luminescence by Vincenzo Casciarolo in 1602. Their commercial emergence as a new class of materials is however often traced to the research of S. Donald Stookey at Corning Glass, who in the early 1960s aimed to develop a glass that turns opaque automatically when exposed to light and vice versa increase its transparency when the light source is withdrawn. By adding silver and copper halides into the glass formula Stookey succeeded in creating a photochromic glass, which to date still forms the basis for light-sensitive glass with self-adjusting transparencies (Geiser 2001, p. 249).

Table 2.1 Stimulus response matrix for a selection of smart materials

Response → ↓ Stimulus	Electrical	Magnetic	Optical	Thermal	Mechanical	Chemical
Electrical		Magnetoelectronics Spinelectronics Spintronics	Electrochromic Electroluminescent Electrooptic Piezochromic Kerr Effect Pockel Effect	Thermoelectric (Peltier)	Piezoelectric Electrostrictive Electrorheological Electrokinetic	Electrolysis Electrochemical Bioelectric Electromigration
Magnetic	Magnetoelectronics Spinelectronics Spintronics, Hall Effect		Magnetooptic Piezochromic	Magnetothermal	Magnetostrictive Magnetorheological	Nuclear magnetic resonance Magnetochemical
Optical	Photoconductive	Optomagnetic	Optical stability	Photothermic	Optomechanical Photoacoustic	Photochemical Photosynthesis Photocatalyst
Thermal	Thermoelectric Superconductivity Radiometer Effect Pyroelectric	Curie point	Thermochromic Thermoluminescent		Shape-memory	
Mechanical	Piezoelectric Electrostrictive	Magnetostrictive	Mechanochromic, Rheochromic		Rheopexic Auxetic Shear-thinning Dilatants Non-Newtonian Pseudo Plastic	
Chemical		Magnetochemical	Color change Litmus Luminescence	Exothermic Endothermic		Catalysis

Adapted from Darrell Mann: Smart Materials Solve Contradictions: paper first presented at ETRIA TRIZ Future Conference, November 2008

Other innovations followed rapidly with one of the most notable being the shape-memory effect found in Nickel-Titanium alloys in 1962 at the now disestablished Naval Ordnance Laboratory (NOL) in White Oak, Maryland, creating a material that is accordingly called Nitinol. The shape-memory effect has since then been discovered in a few other alloys, Nitinol however remains the most popular. Shape-memory alloys (SMA) have the ability to remember the shape in which they were annealed and upon heating try to remake that shape, creating relatively large forces in the process. Starting in the 1980s and early 1990s, several companies are now producing and distributing Nitinol materials and components including connectors, heat engines, and various types of actuators (Cai 2003, p. 9).

Due to the continuous emergence of further types of smart materials the United States army organized a workshop on *Smart Materials, Structures, and Mathematical Issues*, in September 1988 aiming to "identify recent significant developments and breakthroughs in science and technology." Its main focus was to develop a general agreement on the definition and characteristics of a "smart/intelligent material or structure," since the use of different, often complementary terms, such as smart, intelligent, active, multifunctional, or adaptive lead to growing confusion among the scientific community. Dr. lqbal Ahmad, director of the Materials Science Division of the United States Army Research Office and chairman of the workshop, proposes to stick to the term smart and define it technically as:

> A system or a material which has built-in or intrinsic sensor(s), actuator(s) and control mechanism(s) whereby it is capable of sensing a stimulus, responding to it in a predetermined manner and extent, in a short/appropriate time and reverting to its original state as soon as the stimulus is removed (Ahmad 1988, pp. 1, 4).

The Defense Advanced Research Projects Agency (DARPA) in the United States is highly convinced that smart materials will change the capabilities of military systems and commercial applications in the near future. Since the early 1990s they have put serious efforts in the development of advanced actuator designs to move developing concepts of smart materials into practical applications. To them the term smart materials has thus always been used to describe a materials system "constructed of actuation materials, structural materials, structure design, sensing, and control" (Wax et al. 2003, pp. 17–23).

A similar definition is the base of a Foresight study performed by the Institute of Materials, Minerals and Mining, London entitled *Smart Materials for the twenty first century,* which highlights the "development of products with increasing levels of functionality" as a "key to twenty-first century competitive advantage."

> Here, we define 'Smart Materials' as materials that form part of a smart structural system that has the capability to sense its environment and the effects thereof and, if truly smart, to respond to that external stimulus via an active control mechanism (Institute of Materials, Minerals and Mining 2003, p. 9).

To disperse misunderstanding in relation to different terminologies and meanings they continue:

> The terms 'smart', 'functional', 'multifunctional' and 'intelligent' are often used interchangeably. This is reasonable, if confusing, for the first three terms but the last almost certainly suggests a degree of consciousness that does not exist in any non-biological system. There is arguably no such thing as a 'smart material' per se - there are only materials that exhibit interesting intrinsic characteristics which can be exploited within systems, or structures that, in turn, can exhibit 'smart' behavior (Institute of Materials, Minerals and Mining 2003, pp. 5, 11).

2.7.4 Concepts for Information Materials Usage in Architecture

Building upon this notion of "materials which can alter their properties or transmit information merely due to electronic or molecular proceedings," the British architect Mike Davis presented in 1981 the concept of his polyvalent wall, a wall that controls the flow of energy from the exterior to the interior via thin, multifunctional layers. Proposing that this would become the future envelope of a building, removing "the distinction between solid and transparent" he proclaims the shift from "the mechanical age to a 'solid state' era" (Davis 1981, pp. 55–57).

With no lesser conviction emphasizes John Orton in his book *Semiconductors and the Information Revolution: Magic Crystals that made IT Happen*, the utter importance and ubiquitous presence of semiconductor technologies in our current time and grants that they should "rank alongside the Beethoven Symphonies, Concord, Impressionism, medieval cathedrals, and Burgundy wines and we should be equally proud of it" (Orton 2009, p. vi). Similarly claims the Physicist, Karl Wolfgang Böer that:

> Semiconductors have sparked the beginning of a new material epoch. Technology has evolved from the stone age through the bronze and iron ages into the age of semiconductors, materials that are influencing culture and civilization to an unprecedented degree (Böer 2002, p. 2).

While it is a bit simple to reason from the influence of stone and iron on architecture that semiconductors and as such *information materials* will equally and as importantly find their way into the (built) environment they do undoubtedly offer an immensely broad scope of potential applications. These range from the (imaginative) creation of fully transformable immersive environments (something that Winy Maas explored in his 'Barbapapa' and 'Transformer' studios held at ETH Zurich and TU Delft) over climate responsive facade systems and building envelopes, like Decker Yeadon's homeostatic facade system (Decker 2014, p. 78), to novel forms of interaction with and through a dynamic materiality, like Stefan Ulrich's Funktionide, a shape-shifting, soft, and amorphous robot (Dezeen 2009). A more practical and applied suggestion is Manuel DeLanda's reference to the self-monitoring and self-healing capacities of biological structures when emphasizing the potential of such materials to respond to the occurrence of cracks in load-bearing components. He proposes to enhance concrete columns with optical fibers or piezoelectric crystals to analyze incipient fissures, which could consequently be closed by activating embedded shape-memory alloys (DeLanda 2006, pp. 122–123). The earlier mentioned Foresight study provides similar suggestions, emphasizing the advance of 'smart' buildings and infrastructure:

> The report specifically identifies as exemplars the use of embedded sensors to remotely monitor building performance, the use of smart technologies for identity, data collection and management, and decorating paint that electronically changes colour and warns of stress points (Institute of Materials, Minerals and Mining 2003, p. 23).

Although the number of such practical, yet still mostly theoretical, ideas is relatively large they have so far only seldom gone beyond a prototypical state toward more applied and spatial scales. Thus the amount of realized architectural projects challenging the usage of *information materials* and proposing alternatives to existing structures is rather limited. This is due to certainly the low-cost culture in respect to building materials but probably also because of a general lack of knowledge as well as limited access to architectural *information material* products. Michelle Addington and Daniel Schodek add that "materials continue to be chosen not so much for how they perform, but what they connote." In that respect *information materials*, whose fundamental properties are of behavioral nature, are largely overlooked since they possess only little connotative qualities. Departing from this they demand to "move beyond fetishization of the gadgets, and get over our preoccupation with showing off the advanced materials in a purely provocative manner" (Addington and Schodek, pp. 201–203).

Most of the early works employing *information materials* in a spatial context originate however exactly from such an intention, displaying smart materials for their artistic, surprising effects. One of the earliest examples is Sigmar Polke's Thermowand installation at the Musée d'Art Moderne de la Ville de Paris from 1988. Using three different types of thermochromic liquid crystal substances the German artist covered a convex-curved wall with a large area of temperature-sensitive paint, which visualized the daily path of the sun by changing its color in response to the sunlight shining upon it (Ritter 2007, p. 86).

In 1999 the Dutch architecture firm OMA was commissioned by Prada to design a number of flagship stores around the globe. The New York 'Epicenter' opened in 2001 and on top of its avant-garde design it features a number of experimental technologies and new materials. The doors of the changing rooms, for example, are made of Privalite glass, a liquid crystal composite that switches from transparent to translucent when customers enter (Vegesack and Eisenbrand 2006, p. 152).

Another innovative concept is SmartWrap, a proposal of a multifunctional building skin developed by Kieran Timberlake Associates, which bears striking similarities to Mike Davis' polyvalent wall described earlier. The two-layer facade skin is made of transparent, elastic PET foil. Organic photovoltaic cells are printed onto the outer surface to harvest energy, which is stored in thin film batteries and distributed throughout the system through conductive, printed circuits and organic thin film transistors. The foil is further equipped with polymer-based **OLEDs** for lighting and electronic displays, and with chromatic solar protection for adjusting the transmission of light and heat. The inner skin contains pockets of aerogels for insulation and phase-change materials for heat storage. A small portion of the building envelope was demonstrated in August 2003 at the SOLOS exhibition at the Cooper Hewitt National Design Museum (Ritter 2007, p. 140).

Phase-change materials have also been used by the Swiss architect Dietrich Schwarz in the design of his Senior Citizens' Apartments building in Domat/Ems, 2004 Fig. 2.19. The 148 m^2 southern glass facade of the building is made of a latent heat-storing insulation glass, which employs a light-directing prism panel to reflect direct light radiation during summer but permit light penetration in the winter. The low-angled light then passes through the glazing where it hits salt hydrate panels on the inside of the structure and gets stored by melting the material. When the room temperature falls below 26 °C the salt hydrate crystallizes and releases its energy back into the room. When charged the opaque panels turn translucent, making the heat-storing effect visible (Schröpfer and Carpenter 2011, p. 168).

Fig. 2.19 The retirement community in Domat Ems, Switzerland by Schwarz Architekten integrates phase-change materials for latent heat storage within its glazed facade (Kretzer 2015)

On occasion of the international building exhibition IBA in Hamburg four Smart Material Houses have been built in 2013. The model homes are ought to demonstrate new technological approaches for creating more sustainable and energy efficient buildings. The project BIQ by Splitterwerk, Graz incorporates a bioreactor glass facade, which cultivates green microalgae to produce energy and control light radiation and shading. Zillerplus Architekten from Munich created a building that has phase-change materials embedded into its external skin and is able to produce more energy than its residents require. Woodcube by the architektenagentur, Stuttgart is a five-story apartment building that consists almost entirely of wood and demonstrates how traditional techniques can be reinterpreted in a modern way. And the Soft House by Kennedy & Violich Architecture uses photovoltaic cells incorporated into a dynamic textile facade that turns toward the sunlight to harness energy (International Building Exhibition IBA Hamburg).

2.7.5 Material Behavior in Regards to Energy, Time, and Space

Evaluating the above-described examples, it is interesting to observe that while the earlier, more artistic ones focus on emphasizing particular material effects for their entertaining value, the latter, increasingly architectural works deal mainly with energy related issues. Affordance and affordability thus seem to be key aspects for the rise and usage of any smart technology either by in itself being extremely economical or by adding extra value through increased performance, including additional safety or long-term cost reduction in regards to maintenance. Thomas Schröpfer argues that "with buildings and their associated systems contributing [...] an average of 32 % of all greenhouse gases in the US" architects have a certain responsibility to address and prioritize "issues of sustainability, energy efficiency, and material lifecycles" (Schröpfer and Carpenter 2011, p. 178). Yet, as textile designer and researcher Aurélie Mossé points out, that while "sustainability is predominantly understood as a practice aimed at reducing and minimizing the impacts of human actions on the environment," it is "not only an issue of space and matter, but more fundamentally a temporal aspect," and should include an understanding of what to sustain, in whose interest, and for how long it is to be sustained (Mossé 2014, p. 88).

Considering time an essential characteristic of both sustainability and materiality, especially in relation to *information materials*, Sheila Kennedy argues that when designing in the fourth dimension:

> The question for architecture thus becomes not what is a material but when is a material; when does it change from one state to another, and how may its dynamic behavior be designed and experienced in the space of architecture? (Schröpfer and Carpenter 2011, p. 120).

In *Smart Materials and new Technologies* Addington and Schodek note that nowadays many products are labeled green in terms of energy usage. However, the assumption that an accumulation of more powerful products will automatically result in more efficient buildings and settlements and thus less fossil fuel consumption and global greenhouse emissions is misleading since "energy boundaries do not fall into a vertical or horizontal arrangement with a neatly additive accounting system." To them buildings are first of all units of private property, which might eventually be considered "a collection of behaviors that intervene at many different locations in the energy network," but should not be seen as energy systems or as containers for energy systems (Addington and Schodek 2005, p. 220).

Ludger Hovestadt vigorously criticizes the current fixation with terms like sustainability and energy efficiency and proposes to (intellectually) liberate ourselves from energetic constraints and scarcities. He believes that through our knowledge on how to, for example, produce functional materials like solar cells and thus harvest electricity from sunlight, we have in principle access to an abundant source of continuous supply, providing us with "roughly 10,000 times our overall energy consumption today." To him energy is thus not about declining resources, nor about inefficient technology, and certainly not about sustainability, but simply a matter of the unrestricted and far-reaching distribution of knowledge and intellect (Hovestadt 2014, p. 64).

Together with Vera Bühlmann in their book *Printed Physics—Metalitikum I*, they highlight yet another essential aspect of *information materials*, which often tends to be overlooked: the novelty of their fabrication process "using printing technologies." Drawing direct analogies to the revolutionary importance of Gutenberg's printing press in the fifteenth century, which "promoted the secularization of mental horizons in philosophy and modern science," they argue that we are now witnessing an equally important moment in time, initiating the "secularization of a naturalized rationality principle." Just as the invention of the printing process made formerly inaccessible text available as a descriptive medium to society and essentially reformed dominant belief structures, brings "information-technological printing technology" unprecedented new possibilities that might reach far beyond general anticipations.

Bühlmann and Hovestadt build their thesis upon two major lines of argumentation. The first is based on the conception that information technology is profoundly distinct from both matter and energy, since unlike mechanics "it controls the physical conditions symbolically."

The second statement emphasizes the industrial production of such 'symbolic physics' which thus are able to quickly reach a substantial mass, powerful enough to have global effects (Bühlmann and Hovestadt 2013, pp. 12–15). Referring to both the vast impact of Google as well as cellular phone technology on society Hovestadt exemplifies a number of decisive digitally enabled novelties, which are all based upon similar information technological phenomena. These include **microelectromechanical systems** (MEMS), piezoelectronics, semiconductors (like LEDs or OLEDs), but most of all photovoltaics and their potential for an abundance of clean energy (Hovestadt 2013, pp. 60–67).

Such an assumption, challenging not only some of the core principles of contemporary architectural design but our present attitude toward the planet and nature in its entirety, might be hard to simply accept and is in its practical implementation still far from realization. Yet its theoretical foundation provides an immensely powerful base to liberate oneself from the externally imposed burdens of environmentally friendly and ecologically aware design politics and allows the designer to explore *information materials* in relation to other fundamental aspects of architecture, such as space, time, and particularly people (Kolarevic 2014, p. 150). Moreover, the hypothesis of an abundant availability of not only resources but especially possibilities poses extremely challenging tasks and uncovers our current impotence in addressing far-reaching value and context, which are often concealed behind formal aesthetics and eye-catching effects.

Addington and Schodek expand upon this issue and emphasize the responsibility of architects and designers in purposefully integrating novel technologies into our environment. As basic guidelines they formulate six key aspects in relation to energy theory and the essentials of material structure:

- Energy is about motion, and motion can only occur if there is a difference in states between a system and its surroundings.
- The exchange of energy can only take place at the boundary between a system and its surroundings.
- Energy must be accounted for during exchange processes. Any energy exchange that is not 100 % efficient will produce heat. As such, all real world processes produce excess heat.
- Usable energy is lost in every exchange. When there is an energy input in one form, the usable energy output is always lower.
- Material properties are determined by either molecular structure or microstructure. Any change in a material property, such as what happens in a smart material, can only occur if there is a change in one of the two structures.
- Change can only occur through the exchange of energy, and that energy must act at the scale of structure that determines the material property (Addington and Schodek 2005, pp. 221–222).

Building upon this Addington and Schodek declare that any kind of material behavior can be understood by relating to these basic principles, which will empower architects to focus on phenomena and environments rather than material artifacts. Calling for the active exchange of knowledge with other disciplines they highlight the potential of *information materials* as being more than just an exciting new technology but a powerful impulse and chance to question the status quo. This will allow architects to depart from established models, beyond "the notions of efficiency and expediency" and create results that "are not buildings or urban infrastructure, but places of human interaction" (Addington and Schodek 2005, p. 227).

2.8 Final Summary and Conclusion

The present chapter sets forth to investigate the meaning of materiality and how the role of materials in architecture changed and evolved over the course of time. The first part of the chapter is split into the topics *Natural Materials*, *Industrial Materials*, and *Synthetic Materials*, while the second part covers the areas *Digital Materials* and *Information Materials*.

In comparison to the first part, the latter obviously has a much smaller theoretical foundation, spanning a little more than two or three decades, yet in both cases the chosen positions are not intended to reflect a general tendency but rather specific opinions within a diverse context. While the phenomenon of digital materials in regards to freeform or nonstandard architecture is by now largely accepted and publicly renown they do remain exceptional cases and the majority of practices still designs and builds in a fairly traditional sense. In that context it is interesting to observe that the very inventor of CAD systems, Ivan Sutherland, proclaims that using digital tools is only helpful if the result is superior to what can be done by hand.[3] Obviously computer-aided drawing has greatly sped up the design process and being able to copy–paste or delete elements on the fly was not as easy before, yet it was possible. In that sense it remains questionable where the true benefits of these new technologies lie and how to adequately assess them. Robert Aish from Autodesk Research transfers the responsibility to the user and argues:

> A creative tool is one that facilitates [...] customization and can be used beyond what was envisaged by the original tool builder.

He encourages us to challenge both the instruments and ourselves in a critical and reflective way (Sheil and Glynn 2011, p. 11). In relation to *information materials* the reluctance of the discipline is even higher, mostly because the technology still seems too far off in order to be considered serious. However, there are other factors such as scalability, longevity, costs, sustainability etc., as well as the fact that the materials are usually designed and developed for particular situations and purposes, which mutually hinder the progressive advancement in an architectural context.

In both cases a closer cross-disciplinary exchange could prove helpful, especially since much of the technological development happens in other sectors, such as the aerospace, automotive, and military industries. Moreover, despite the time it takes to create and establish new technologies on the market so they can be promoted to other areas, it might be the very standards and norms, which emerged during Industrialization, that impede their architectural application and which might require reassessment in respect to the current era.

What however remains striking and highly motivating is the observation of an underlying desire for the creation of more dynamic, more fluid, flexible, or liquid spaces, which seems to have (re)emerged with the occurrence of digital technologies around the turn of the century. While these formulations were in the beginning mostly tied to virtual environments, akin to Markos Novak's liquid architecture, forerunners in digital fabrication like Greg Lynn manifest them physically, and visionaries such as Kas Oosterhuis even attempt to maintain the dynamicity of the digital design process through the creation of (inter)active spaces. As such it remains very likely that both digital materials as well as *information materials* will influence architecture on a variety of levels and scales. Yet considering the duration of a building's construction and the time it is ought to last, as well as its tightly tied cultural and social implications, meaning, and value it is very possible that change will not emerge from within the architectural discipline but rather through user demands, applications, and products. Often such shifts are much more subliminal than initially anticipated. While plastics, for example, did not, or only for a short period, lead to the architectural revolution its supporters proclaimed, the material is more than omnipresent today. In a similar sense might *information materials*, at least for the time being, be much more powerful in smaller scales emerging from particular consumer related demands.

In that sense a reformulation of the architectural practice and the development of new instruments will be the keys in order to remain up-to-date. Concluding from such a demand a fundamental evolution of architecture both

[3]See Sect. 2.6.1 *The Emergence of Digital Design and Fabrication Techniques* for more details.

topologically and formally might however still require decades of integrative research and collaboration. What can be hoped for in the nearer future is a more comprehensive and less hierarchical exchange among adjacent disciplines and thus a smoother integration of technologies into space and structure. This will require architects to (re)formulate contexts and visions, within which they position their work both in terms of origin and perspective, rather than just updating existing systems with new technologies or celebrating them for the sake of their spectacularity.

In order to provide a fertile ground for such explorations it is necessary to develop or use methods for the dissemination and access to knowledge that feel appropriate and contemporary. While many of the described treatises in this chapter were intended as references for education and training, relating to the then state-of-the-art, today's material world is much more complex and multifaceted for to be approached in such a didactic manner. Retaining the necessary broadness while avoiding superficiality, yet providing depth and credibility will be the key to create a general awareness of related technological progress among the architectural discipline. The following chapter will thus set out to explore novel ways for knowledge mediation and teaching in order to incept such understanding early on during an architect's career.

References

Ackerman JS (1997) Villard de Honnecourt's drawings of Reims Cathedral: a study in architectural representation. Artibus et Historiae 18(35):41–49

Addington M, Schodek DL (2005) Smart materials and new technologies for the architecture and design professions. Architectural, Oxford

Ahmad I (1988) Smart structures and materials. In: Rogers CA (ed) Proceedings of U.S. Army Research Office workshop on smart materials, structures and mathematical issues. Virginia Polytechnic Institute & State University, Technomic Publishing Co., Inc, pp 13–16

Alberti LB (1988) On the art of building in ten books (trans: Rykwert J, Leach N, Tavenor R). MIT Press, Cambridge

Ashby MF, Ferreira PJ, Schodek DL (2009) Nanomaterials, nanotechnologies and design: an introduction for engineers and architects. Butterworth-Heinemann, Amsterdam

Ayres P, Tamke M, Thomsen MR (2012) Digital material. Riverside Architectural Press, Toronto

Banham R (1980) Theory and design in the first machine age, 2nd edn. The MIT Press, Cambridge

Bell VB, Rand P (2006) Materials for design. Princeton Architectural Press, New York

Beorkrem C (2013) Material strategies in digital fabrication. Routledge, New York

Böer KW (2002) Survey of semiconductor physics, 1, electrons and other particles in semiconductors. Wiley, New York

Bogner D, Peter N (2001) Frederick J. Kiesler: endless space. Hatje Cantz, Ostfildern

Borden I, Rendell J (2000) Intersections: architectural histories and critical theories. Routledge, London

Boucher B (1998) Italian baroque sculpture. Thames and Hudson, New York

Braham A (1980) The architecture of the French enlightenment. University of California Press, Berkeley

Braham WW, Hale JA, Sadar JS (2007) Rethinking technology: a reader in architectural theory. Routledge, London

Buchanan P (2011) The autopoiesis of architecture dissected, discussed and decoded. Archit Rev

Bühlmann V, Hovestadt L (2013) Introduction—printed physics. In: Hovestadt L, Bühlmann V (eds) Printed physics—Metalithikum I. Ambra V, Vienna, pp 11–16

Cache B (1995) Earth moves: the furnishing of territories. The MIT Press, Cambridge

Cai CS (2003) Applications of smart materials in structural engineering. Louisiana Transportation Research Center, Baton Rouge

Caneparo L (2013) Digital fabrication in architecture, engineering and construction. Springer, Dordrecht

Carus TL, Bailey C (1921) On the nature of things (trans: Bailey C), pp 488–499. Clarendon Press, Oxford

Corbusier L (1964) When the Cathedrals were white, a journey to the country of timid people. McGraw-Hill, New York

Corbusier L (1970) Five points towards a new architecture. In: Conrads U (ed) Programs and manifestoes on 20th-century architecture. MIT Press, Cambridge, pp 99–101

Corbusier L (1986) Towards a new architecture. Dover Publications, New York

Corser R (2010) Fabricating architecture: selected readings in digital design and manufacturing. Papress, New York

Davis M (1981) A wall for all seasons. RIBA J 88(2):55–57

Decker M (2014) Adapting matter. In: Kretzer M, Hovestadt L (eds) ALIVE: advancements in adaptive architecture. Birkhäuser, Basel, pp 78–81

DeLanda M (2001) Philosophies of design: the case of modelling software. In: Salazar J (ed) Verb processing. Actar, Barcelona, pp 131–142

DeLanda M (2004) Material complexity. In: Leach N, Turnbull D, Williams C (eds) Digital tectonics. Wiley, Chichester, pp 14–21

DeLanda M (2006) Smart materials. Domus 894(07):122–123

Dezeen (2009) Funktionide by Stefan Ulrich. http://www.dezeen.com/2009/10/05/funktionide-by-stefan-ulrich/. Accessed 15 Mar 2015

Dolphijn R, van der Tuin I (2012) New materialism interviews & cartographies. Open Humanities Press, Ann Arbor

Durand J-N-L (1831a) *Abriss der Vorlesungen über Baukunst, gehalten an der Königlichen Polytechnischen Schule zu Paris: Erster Band.* Werdersche Kunst- und Buchhandlung, Carlsruhe und Freiburg

Durand J-N-L (1831b) *Abriss der Vorlesungen über Baukunst, gehalten an der Königlichen Polytechnischen Schule zu Paris: Zweiter Band.* Werdersche Kunst- und Buchhandlung, Carlsruhe und Freiburg

Evers B, Thoenes C (2003) Architectural theory: from the renaissance to the present. Taschen, Köln

Fernandez J (2006) Material architecture: emergent materials for innovative buildings and ecological construction. Architectural Press, Boston

Foresight Smart Materials Taskforce (2003) Smart materials for the 21st century. Institute of Materials, Minerals and Mining, Foresight Materials Panel

Forty A (2012) Concrete and culture: a material history. Reaktion, London

Friedman Y (2006) Pro domo. Actar, Barcelona

Fuller RB (1969) Operating manual for spaceship earth. Southern Illinois University Press, Carbondale

Geiser K (2001) Materials matter: toward a sustainable materials policy. MIT Press, Cambridge

Gershenfeld N (2012) How to make almost anything: the digital fabrication revolution. Foreign Aff 91(6):43–57

Giedion S (1971) Architecture and the phenomena of transition: the three space conceptions in architecture. Harvard University Press, Cambridge

Goldstein SC, Mowry TC (2004) Claytronics: an instance of programmable matter. Wild and crazy ideas session of ASPLOS

Goldstein SC, Campbell JD, Mowry TC (2005) Invisible computing—programmable matter. Computer 38 (6):99–101

Gramazio F, Kohler M (2008) Digital materiality in architecture. Lars Müller Publishers, Baden

Guin A (2012) Programmable matter—claytronics. Paper presented at the presented at the 58th international instrumentation symposium, San Diego, California, 4–8 June 2012

Haque U (2007) The architectural relevance of Gordon Pask. Archit Des 77(4): 4. Dsocial: Interactive Design Environments 54–61. doi:10.1002/ad.487

Hensel M, Menges A (2008) Inclusive performance: efficiency versus effectiveness towards a morpho-ecological approach for design. Archit Des 78 (2). Versatility Vicissitude 54–63. doi:10.1002/ad.642

Hill J (2006) Immaterial architecture. Routledge, London

Hovestadt L (2009) Jenseits des Rasters: Architektur und Informationstechnologie/Beyond the Grid: Architecture and Information Technology. Birkhäuser, Basel

Hovestadt L (2013) A fantastic genealogy of the printable. In: Hovestadt L (ed) Printed physics—Metalithikum I, and Vera Bühlmann. Ambra V, Vienna, pp 17–70

Hovestadt L (2014) Domesticating a world of printed physics. In: Kretzer Manuel, Hovestadt Ludger (eds) ALIVE: advancements in adaptive architecture. Birkhäuser, Basel, pp 62–71

Http (2015) http://www.architectural-review.com/essays/the-autopoiesis-of-architecture-dissected-discussed-and-decoded/8612164.article. Accessed 18 Mar 2015

International (2015) International building exhibition IBA Hamburg. Smart building materials for the future: smart material houses. http://www.iba-hamburg.de/en/projects/the-building-exhibition-within-the-building-exhibition/smart-material-houses/projekt/smart-material-houses.html. Accessed 18 Mar 2015

Iwamoto L (2009) Digital fabrications: architectural and material techniques. Princeton Architectural Press, New York

Jeska S (2008) Transparent plastics: design and technology. Birkhäuser, Basel

Johnson P (1960) Three architects. Art America 70–75

Kahn L (2003) In: Twombly RC (ed) Louis Kahn: essential texts. W.W. Norton & Company, New York

Kennedy S (2011) Responsive materials. In: Schröpfer T, Carpenter J (eds) Material design informing architecture by materiality. De Gruyter, Berlin, pp 118–131

Kidson P, Fernie EC, Crossley P (1990) Medieval architecture and its intellectual context: studies in honour of Peter Kidson. Hambledon Press, London

Kiesler F (1925) Vitalbau-Raumstadt-Funktionelle Architektur. De Stijl 6(10–11):141–146

Kolarevic B (2003) Architecture in the digital age: design and manufacturing. Spon Press, New York

Kolarevic B (2014) Adaptive architecture: low-tech, high-tech, or both? In: Kretzer M, Hovestadt L (eds) ALIVE: advancements in adaptive architecture. Birkhäuser, Basel, pp 148–157

Koolhaas R (1994) Delirious New York: a retroactive manifesto for Manhattan. Monacelli Press, New York
Kruft H-W (1994) A history of architectural theory: from Vitruvius to the present. Zwemmer, London
Laugier M-A (1977) An essay on architecture. Hennessey & Ingalls, Los Angeles
Lavin S (2014) Neither plastic as concept, plastic as material. In: Bell M, Buckley C (eds) Permanent change: plastics in architecture and engineering. Princeton Architectural Press, New York, pp 15–20
Leslie T (2013) Chicago Skyscrapers, 1871–1934. University of Illinois Press, Urbana
Lynn G (1993) Architectural curvilinearity: the folded, the pliant and the supple. Archit Des Fold Archit 63(3–4):24–31 (Academy Editions)
Mann D (2009) Smart materials solve contradictions. Triz J. http://www.triz-journal.com/smart-materials-solve-contradictions. Accessed 19 Mar 2015
Menges A (2012) Material computation: higher integration in morphogenetic design. Archit Des 82(2). Mater Comput 14–21. doi:10.1002/ad.1374
Mossé A (2014) Smart materials: designing a timescape of interconnectivity for more resilient practices of inhabitation. In: Kretzer M, Hovestadt L (eds) ALIVE: advancements in adaptive architecture. Birkhäuser, Basel, pp 86–91
Muecke M (2005) Gottfried Semper in Zurich: an intersection of theory and practice. Culicidae Press, Ames
Mueller RK (1955) For the plastics industry. In: Plastics in building; the uses, past and present, and the potentialities of plastics in building as reported at a conference conducted by the Building Research Institute, October 27 and 28, 1954, at the Chamber of Commerce of the United States in Washington, DC, pp 127–130. National Research Council (U.S.), Washington DC
Orton J (2009) Semiconductors and the information revolution: magic crystals that made IT happen. Academic Press, Amsterdam
Oxman N (2010a) Material-based design computation. PhD dissertation, Massachusetts Institute of Technology
Oxman N (2010b) Structuring materiality: design fabrication of heterogeneous materials. Archit Des 80 (4):78–85. doi:10.1002/ad.1110 (The New Structuralism: Design, Engineering and Architectural Technologie*s*)
Pérez-Gómez A (2005) From treatise to story: the changing nature of architectural discourse from the renaissance to the eighteenth century. SPELL: Swiss Pap English Lang Lit 17:35–49. doi:http://dx.doi.org/10.5169/seals-100019
Perrault C (1993) Ordonnance for the five kinds of columns after the method of the ancients. Getty Center for the History of Art and the Humanities, Santa Monica
Ritter A (2007) Smart materials in architecture, interior architecture and design. Birkhäuser, Basel
Sadler S (2005) Archigram: architecture without architecture. The MIT Press, Cambridge
Schindler C (2009) Ein architektonisches Periodisierungsmodell anhand fertigungstechnischer Kriterien, dargestellt am Beispiel des Holzbaus. PhD dissertation, Swiss Federal Institute of Technology (ETH), Zurich
Schröpfer T, Carpenter J (2011) Material design: informing architecture by materiality. De Gruyter, Berlin
Schumacher P (2011) The autopoiesis of architecture: a new agenda for architecture, vol I. Wiley, Chichester
Scott FDE (2008) Living archive 7: ant farm—allegorical time warp: the media fallout of July 21, 1969. Actar, Barcelona
Semper G (1851) Die Vier Elemente der Baukunst. Friedrich Vieweg und Sohn, Braunschweig
Semper G (1852) Wissenschaft, Industrie und Kunst: Vorschläge zur Anregung Nationalen Kunstgefühles bei dem Schlusse der Londoner Industrie-Ausstellung. Friedrich Vieweg und Sohn, Braunschweig
Semper G (1878) Der Stil in den technischen und tektonischen Künsten, Erster Band. Friedrich Bruckmann's Verlag, München
Sheil B, Glynn R (2011) Fabricate: making digital architecture. Riverside Architectural Press, Toronto
Speaks M (1995) Folding toward a new architecture. In: Earth moves: the furnishing of territories, Bernard Cache, xiii–xix. The MIT Press, Cambridge
Spiller N (2008) Digital architecture now: a global survey of emerging talent. Thames & Hudson, London
Sutherland IE (2003) Sketchpad: a man-machine graphical communication system. Technical Report No. 574, University of Cambridge Computer Laboratory
Tibbits S (2014) 4D Printing: multi-material shape change. Archit Des 84(1). High Definition: Zero Tolerance in Design and Production 116–121
Toffoli T, Margolus N (1991) Programmable matter: concepts and realization. Physica D 47(1–2):263–272. doi:10.1016/0167-2789(91)90296-L
Vasari G, Maclehose LS, Baldwin Brown G (1960) Vasari on technique; being the introduction to the three arts of design, architecture, sculpture and painting, prefixed to the lives of the most excellent painters, sculptors, and architects. Dover Publications, New York
Vitruvius Pollio M (1486) English edition: The ten books on architecture (trans: Morgan MH) (1914). Harvard University Press, Cambridge
Von Vegesack A, Jochen E (2006) Open house: architecture and technology for intelligent living. Vitra Design Stiftung, Weil am Rhein
Wax S, Fischer G, Sands R (2003) The past, present, and future of DARPA's investment strategy in smart materials. JOM: J Miner Met Mater Soc 55(12):17–23

Weiss W (2010) "The Future Won't Wait" From Monsanto Magazine in 1960. http://www.yesterland.com/futurewontwait.html. Accessed 14 Oct 2016

Wiener N (1961) Cybernetics: or control and communication in the animal and the machine. MIT Press, New York

Wright FL (1936) 1935: organic architecture. Archit J 177–199

Yarsley VE, Couzens EG (1941) Plastics. Penguin Books, Harmondsworth

Zumthor P (2006) Atmospheres: architectural environments, surrounding objects. Birkhäuser, Basel

Zumthor P, Oberli-Turner M, Schelbert C (2006) Thinking architecture. Birkhäuser, Basel, p 2

3 Materiability—An Attempt for the Education of an Information Material Literacy in Respect to Emerging Materials

3.1 Introduction

> The present state of our urban situation cannot be solely addressed by its infrastructure or physical orientation. Our environment has to be understood as an ever-growing organism, influenced by every vector defining our society. New technologies, as well as the overall instant availability of information, not only transform our daily life and routine, but also redefine the notion of the city itself. Recognizing urban structures as process-oriented, nonstatic, fluid environments requires an experimental and all-encompassing architecture that is physical, virtual and dynamic in its very condition.
>
> Hollein (2002, Architectural Laboratories, p. 7)

In the introduction to Greg Lynn's and Hani Rashid's *Architectural Laboratories*, curator and architect Max Hollein, emphasizes the ever-changing and evolving nature of our environment in respect to continuous technological advancement and architecture's responsibility to address such developments in a vital and progressive manner. Just as the role and tasks of the architect will continue to change, he highlights the necessity for architectural education to prepare students for upcoming challenges and uncertainties. This not only involves the provision of the necessary technical equipment and training but especially the development of new relationships between the various involved areas and the establishment of alternative solutions for knowledge mediation and teaching. In a similar manner demands Victoria Ballard Bell, author of *Materials for Design*, particularly in relation to material culture, a more holistic and integrative approach. She criticizes the general exclusion of material-related issues from design-centered education and highlights an overlaying misconception, which often results in materials being chosen at the very end of the process, "as if they are a mere afterthought." Quite the contrary she asserts that due to continuously advancing fabrication and processing methods for the development of new materials, or the improvement of existing ones, materiality deserves increased attention early on during an architect's training. She refers to a report by Ernest Boyer and Lee Mitgang, entitled *Building Community: A new Future for Architecture Education and Practice*, which, among other essential educational goals, demands courses that jointly communicate technical and practical as well as theoretic and artistic skills. Likewise, she argues for an early conveyance of a basic but broad understanding of materiality and the implementation of new didactic prototypes and pedagogic methodologies that foster this approach (Ballard Bell and Rand 2014, pp. 9–10).

In *Architecture 3.0—The Disruptive Design Practice Handbook*, Cliff Moser emphasizes that both the building recession with the concomitant decline of the construction industry but also architectural education's sole concentration on growing experts in 'design for building' has

M. Kretzer, *Information Materials*, DOI 10.1007/978-3-319-35150-6_3

resulted in isolated and overspecialized individuals who find it hard to thrive in a dwindling market. Instead he claims that architecture schools should teach students how to think broadly and aim at building a profession that focuses on finding design "solutions leveraged with collaboration and networking" (Moser 2014, pp. 15–17). Nic Clear, head of architecture and landscape at the University of Greenwich, expands upon the importance of an architect's diverse knowledge and skills, which should encompass a wide range of related disciplines. Multi- and interdisciplinary ways of working, already during education, are, according to him, key to develop an open understanding of collective power that overcomes the anachronistic notion of the architect representing the lead professional. Extending a school's range of learning possibilities beyond the classical set of skills is thus crucial, not only because "the majority of students entering architectural education will not become professional architects," but also because the academic environment represents a place that should encourage speculation and freedom and thus the creation of new talents and ideas. To Clear the core task of architectural education hence lies in empowering students to think, develop, illustrate, mediate, and execute spatial ideas, whether they end up becoming buildings or not (Clear 2014, pp. 93–94, 99–101). Thomas Fisher, Dean of the College of Design at the University of Minnesota, also sees the main tasks of architectural schools in raising confident personalities who learn to think critically while retaining an open perspective and unprejudiced mind. He describes the transformation we're currently experiencing as

> A series of shifts from a mechanistic worldview view to one of organic flows; from an urge to dominate nature to one that seeks a balance with it; from mass production to mass customization; from large bureaucratic organizations to smaller project-based operations; from specialized jobs to versatility; and from professional autonomy to participatory teamwork (Fisher 2000, p. 1).

In this context, he believes that design will play as essential a role as science and technology have during the Industrial Revolution. He claims that key to such progress will be based on interdisciplinary and flexible models, which will not only allow solving upcoming challenges much faster and more efficient but also encourage a reconnection between practice and education due to a more versatile common understanding and knowledge (Fisher 2000, pp. 5, 46–49).

New York-based architects Nanako Umemoto and Jesse Reiser consent that both in academic and industrial environments "truly innovative projects are not solitary productions, but the result of collaborations among the disciplines, each contributing its own expertise" (Unemoto and Reiser 2014, p. 212). At the base of interdisciplinary work is, however, not only a non-hierarchic structure and a shared interest in approaching a mutual goal but especially the proficiency of all involved parties to communicate and converse, essentially the ability to speak the same language. This becomes particularly evident when areas that reside outside the usual comfort zone of the architect are involved in the discussion, like materials science, biology, or chemistry, the birthplaces of many advanced *information materials*. So, in order to encourage young architects to confidentially exchange across multiple disciplines they need to be provided with possibilities to become literate in a shared way of expression.

The present chapter sets forth to elaborate upon a certain didactic model, which from now on is referred to as the *materiability* approach. The term *materiability* was coined by Prof. Dr. Ludger Hovestadt, highlighting the abundance of possibilities emerging from the synthetic creation of new active materials and the necessity for architects and designers to learn using and qualifying their potential. This definition forms the foundation of a pedagogical attempt for integrative and explorative material education with a particular focus on new, emerging *information materials*. Rather than prescribing a particular method or predefined solution, the focus of *materiability* is on providing open access to information and suggesting a relevant range of possibilities. The students are thus not taught 'what-to-do' but instead encouraged to learn 'how-to-do.' *Materiability* does not prioritize instructing certain techniques and skills or foreordain particular design solutions or material applications but aims at providing a stable and rich framework for independent self-development. This pedagogical idea is encouraged by the above-described tendencies and based upon emerging phenomena on the acquisition of knowledge that can for example be witnessed in the usage of the World Wide Web and especially the growing importance of social media. Since the year 2000, the global Internet usage has grown from 5.1 % of the world population to 34.3 % in 2012 (Internet World Stats 2012), will reach 45 % by 2016 (The Network 2012) and, if one was to believe *Forum of the Future* founder Jonathan Porrit, will be accessible by 97.5 % of humankind in 2050 (Porrit 2013, p. 27). Such a fundamental increase in users will inevitably lead to much more online content and a loss in quality might be an obvious side effect. Yet at least for the time being it seems that especially younger people have found extremely effective ways and methods to quickly and successfully scout through the abundance of digital information and extract meaningful outcome even though they have never had any kind of formal training in its usage. Only through continuous and intuitive application and the resulting experience, and of course a number of very smart algorithms that help providing personalized data based on previous behavior, users learn to abstract and distinguish suitable from insignificant information. The Internet thus does not advocate particular results but provides an unlimited amount of potential solutions, embedded in a stable and consistent framework that facilitates orientation, from which the literate user extracts meaning and chooses the most appropriate and valid options.

The *materiability* method aims to draw from this attitude toward digital and social media and proposes to establish a similar strategy for approaching and mediating *information materials*. The methodology, and consequently the present chapter, is split into three consecutive parts, each providing a certain aspect of the larger concept.

The first part, *The Process of Physical Making,* highlights the importance of hands-on experience and creative making. It elaborates upon artistic strategies on prototyping, exemplified by the creative practice of Kurt Schwitters, Frederick Kiesler, and Ray and Charles Eames, who all share an affinity to focus on the dynamic processes of art and design and not its final outcomes. The relevance of such dynamic models is validated in relation to *information materials*, whose inherent transformative properties require an understanding of behaviors in relation to time and space. The section is concluded with a description of the educational methodology of the *materiability* approach and its core structure employed during physical workshops and student projects.

Part two, *Access to Information,* addresses issues like openness, sharing, and collective progress. It describes three strategies, which aim at disclosing knowledge for the emancipation from established structures, including Stewart Brand's *Whole Earth Catalog*, Yona Friedman's *Ville Spatiale*, and Bernard Rudofsky's *Architecture Without Architects*. The section maps out a number of contemporary approaches for the display and mediation of material knowledge in relation to architecture and design. It is concluded with an explanation of the *materiability research network*, an online platform, material database, and educational community that provides exclusive access to *information materials* and related content.

The third and final part, *Information Material Literacy,* describes the urge to acquire a comprehensive material understanding for the progressive utilization of *information materials* in architecture and design projects. Highlighting that literacy not only empowers people to express themselves and converse in a sophisticated manner, but especially provides them with an unprecedented increase in choices and possibilities, it presents a number of current views on the power of the Internet, social media, and consumer focused digital fabrication tools. Expanding upon their relevance for architectural education the section is finalized with an overview of the social aspects of the *materiability research network*, a website that was initiated to make emerging material developments more accessible to the architecture and design communities.

3.2 The Process of Physical Making

> Every form is the frozen instantaneous picture of a process. Thus a work is a stopping-place on the road of becoming and not the fixed goal. We acknowledge works which contain a system within themselves, a system which has not been evolved before the work started but has evolved in the course of it.
>
> Schwitters (1924, Nasci, p. 351)

During the mid-1920s, the German artist Kurt Schwitters collaborated with a number of former Dadaist and constructivist contemporaries, including El Lissitzky, with whom he formulated practical, scientific principles for the creation of art. His masterpiece was the epic architectural assembly the 'Cathedral of Erotic Misery,' also referred to as 'Merzbau,' which he began around 1923. What started as a purely exploratory experiment, building upon distinctive artistic rules and expressions, transformed over the following twenty years into a surrealistic walk-through sculpture that eventually encompassed all eight rooms of his private apartment in Hannover, Germany. Unfortunately the project was completely destroyed during the Second World War, but even though the knowledge about its true form and extent is limited to a few surviving photographs its influence and meaning for contemporary art and architecture remains exceptional. Schwitters built the spatial installation from randomly found elements and without having any formal training into craftsmanship (Bonnemaison and Eisenbach 2009, p. 17). He did not approach it with a preconceived vision of what it should become, but rather developed it through the particular qualities and properties of the materials and artifacts he incorporated and their inherent interrelations. In that sense Merzbau was never about its finished appearance, but more about the dynamic, constantly evolving, and changing process of its creation (Burns Gamard 2000, p. 5). These continuous modifications spanned through several personal periods of interest and research, different artistic phases, and a variety of spatial explorations and expressions. Some accounts actually indicate that in certain areas the structure was up to six layers deep and incorporated movable walls and columns in order to reveal its insides. So even though the installation changed its physical appearance over the course of time, it still contained its beginnings. Rather than replacing parts that had become obsolete they became ingrown into the larger assembly, materializing his idea of the house as body and the body as house (De Zegher and Butler 2010, p. 169).

Schwitters emphasis on continuity, on the fusion of form and formlessness, materiality and immateriality, reminds strongly of Frederick Kiesler, who around the same time challenged normative ideologies by demanding the development of a new architectural language and especially a change in critical thinking. Kiesler wanted to defy static building structures by physically manifesting the effects of time and motion on space and spatial perception through organically appearing forms, mobile and flexible structures, or illusory environments. Instead of imitating principles found in nature or glorifying technological achievements he aimed at merging humans, architecture, technology, and the environment through a dynamic, fluid, and continuous construction. Through this he wanted to stress the creative and interactive relationships that occur in human situations, just as Schwitters saw 'Merz,' the name he had given to his personal artistic principles, as a spatial and temporal model for architecture and urban planning based on "creating relationships, preferably between all things in the world" (Elger 2004, p. 22).

Thomas Fisher also stresses the importance of correlation, interdependence, and effect, consequently encouraging architectural students to continuously question their own work and evaluate its meaning beyond aesthetic and formal appearances but in respect to a user's happiness and the potential desires of future generations. To Fisher this involves the development of an architectural understanding that does not consider objects solely as finished and isolated entities but rather as constantly evolving and performing series of events. Instead of becoming specialists, trying to keep up with the rising demands of the market, which he believes "discourages good design," he proposes a process-oriented model, exemplified through the creative practice of Ray and Charles Eames, which he thinks will allow to resist commodification.

Fisher explains that for the Eameses design was an exploratory process driven by curiosity and excitement and not the attribute of a finished product. The objects they created were hence by-products of material, fabrication, and design investigations that could have gone into many different directions.

Schwitters' and Kiesler's visions, which not only time-wise but also conceptionally have various similarities as well as the much later approach of the Eameses' share an affinity on focusing on the creative process as a dynamic and ongoing method instead of celebrating the final outcome.

This open and unbiased perspective stands in strong contrast to the disciplinary and highly specialized agencies taught within the modern university as well as prevalent tendencies in architectural practice, which tend "to view design projects as objects or sites to be contemplated in isolation, a mentality evident in the way we often photograph projects devoid of people, removed from their physical context, and frozen in time, usually just after completion," as Fisher implies (Fisher 2000, pp. 10, 31, 44–45, 75). Whilst Fisher uses this argumentation in order to stress a building's dynamic and continuous role in society as a place of performance and meaningful events, thinking and designing in terms of processes and behaviors seems, especially in relation to *information materials,* a viable strategy worthwhile investing in.

3.2.1 The Inherent Dynamics of Information Materials

> Understanding materials also means understanding that all materials move.
>
> Ochsner (2012, p. 6)

As exemplified during the previous chapter all materials are inherently dynamic and continuously behave and respond to external inputs, yet mostly in a way that is not directly perceivable with our human senses (Kretzer 2014, p. 73). *Information materials* however are, capable of responding to environmental influences immediately and in a controlled way. Active behaviors are part of their definition, intrinsically embedded into the materials' microstructures. In order to be able to work with them and apply them sensibly and usefully it is hence necessary to understand these materials in terms of their behavioral phenomena over time and not as fixed elements. This requires both experience in their transformative capabilities as well as a basic idea of their internal processes, including their composition and response to energy fields. One possibility to acquire this knowledge can be to break the materials down into their normative ingredients (assuming that they are artificially created or post-processed), understand the properties of these ingredients, and then physically reassemble the materials. Such experimental exploration helps to comprehend the interplay of the materials' individual elements, fosters their technological assessment, and reveals both their potentials and limitations. This leads to profound knowledge in terms of learning new skills and techniques but also proficiency in finding coherence between assembly procedure and behavioral phenomena. Essentially it helps asserting how changes in fabrication relate to particular material properties, which in return can then be tuned to special demands or requirements that vary from standardized market solutions. The acquisition of such experience requires mainly an honest and candid interest in related procedures but does not have to become deeply sophisticated. At the core of this approach is not the mediation of a specific expertise but rather a basic conception of certain principles, which can then form a crucial backbone, in order to communicate progressively across adjacent disciplines. Such interdisciplinary exchange might invigorate the discovery of formerly unknown applications or even incept the improvement of existing or the development of new materials.

3.2.2 The Educational Framework of the Materiability Approach

The projects and workshops that emerge from the *materiability* approach, such as *ShapeShift*, *Phototropia*, *Animated Textiles,* or *Dynamics in*

Extreme Environments, just to name a few, all share a similar structure at their core, a duality-based system being a combination of both bottom–up and top–down strategies, as shown in Fig. 3.1. This concept has since its inception in 2010 been steadily refined and adapted in relation to context, duration, location, technical infrastructure, background, and experience level of the involved students or workshop participants.

Generally the projects emerge from the discovery of one or multiple interesting *information materials*. The respective materials are then studied in more detail and after acquiring a basic theoretical understanding of their working principle and composition (depending on the complexity of the materials) experts are approached to receive more specific information. This can involve gaining first-hand insights in (industrial) fabrication procedures, solving or improving particular assembly details, understanding tendencies in commercial application, or even collaborating throughout the course of the project and especially during an early introductory workshop.

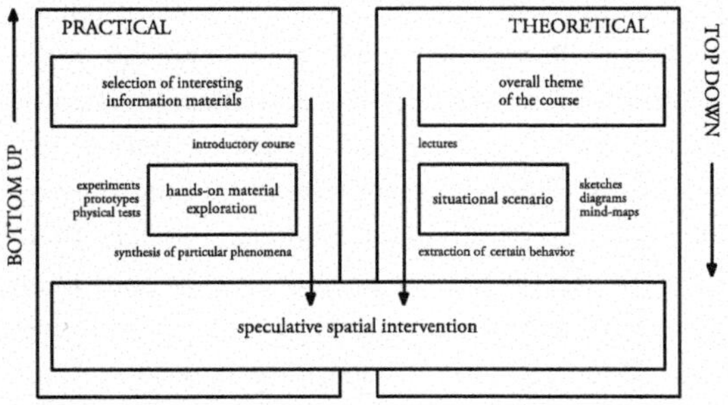

Fig. 3.1 Graphical representation of dualiy in materiability methodology

During these introductory phases, which according to the project scope and available materiality vary in-depth, length, focus and intensity, the students get theoretical information on the materials' functionality and practical instructions on how to self-make them. They are then provided with a hypothetical abundance of resources in order to explore them freely, without having to worry about limited supplies.[1]

By this the students are encouraged to physically experience the functionality of materials, comprehend their working principles, the relationship of fabrication procedures and materials' behavior, and finally extract and emphasize one or more particularly exciting material phenomena.

While the students, who are from the very beginning urged to work in small teams, are gaining confidence in using the materials, they are in parallel asked to think about architectural situations, for the time being without a direct correlation to the available materiality. These situations should be related to the overarching theme of the course and focus on certain human or environmental behaviors, actions, performances, or states, which could be enhanced or improved through some kind of adaptive or responsive system, without actually naming or designing that system yet. Lectures on the topic, often by invited guests who are well established in the respective areas and who might cosupervise parts of the course, selected readings and literature, as well as a vast collection of sample projects help the students to get started within a commonly unknown field. Throughout this part they are animated to think of spaces in terms of dynamic territories, considering design and architecture not as static and immobile objects but as places of continuity that constantly evolve and change over time. They are asked to describe the various stages of a particular spatial situation in terms of flows and behaviors and explore possibilities of visually representing and mediating the dynamic aspects of their findings.

Finally the two approaches, the bottom–up material exploration and the top–down situational scenario, are to be brought together through a speculative spatial intervention. The relation to reality and presence is left open to the students, yet they have to express a certain architectural relevance and are intended to reach a human perceivable scale through either their size or the amount of incorporated components, so they can be understood spatially. This increases the scope and complexity of the

[1]The amount of provided materials depends on the respective material types and associated costs and availability.

projects beyond pure material exploration and prototyping since larger assemblies require very different structural and aesthetic architectures than individual elements do.

If the focus of the *materiability* approach would be on developing particular design answers, the success of a project could be measured by the balance in between the aestheticization of the material phenomena and the effect of the system on the selected situation. However, the intention of *materiability* is not in developing products or proposals but in mediating an understanding of space and materiality as dynamic, fluid entities and in discovering appropriate means for the representation and dissemination of such findings. The final results are thus secondary also since the intention is to avoid associating a particular material prematurely with a certain product or application, which would potentially reduce its capabilities. The outcomes of these workshops and courses should hence be evaluated by the vision they convey and the imaginative spectrum they unclose instead of their practicality.

3.3 Access to Information

> We are as gods and might as well get good at it. So far, remotely done power and glory - as via government, big business, formal education, church - has succeeded to the point where gross defects obscure actual gains. In response to this dilemma and to these gains a realm of intimate, personal power is developing-power of the individual to conduct his own education, find his own inspiration, shape his own environment, and share his adventure with whoever is interested.
>
> Brand (1968, Whole Earth Catalog, p. 2)

In the midst of the 1960s counter-cultural movement, Stewart Brand published the first edition of *The Whole Earth Catalog*, a sixty-four pages long accumulation of essays, articles, and product reviews focusing on self-education, ecology, and low impact living. Brand's prime demand was to democratize and decentralize technology. He wanted to provide unrestricted access to tools and ideas for an anti-authoritarian society that urged to demarcate itself from the status quo. Brand understood the purpose of tools in the processes they facilitate to approach and solve certain tasks, enabling, empowering, and liberating their user (Turner 2006, p. 83). The vast encyclopedic compendium, which quickly grew in popularity and size, turned into a community meeting place in print, connecting and encouraging its readers to advance their independence and autonomy. By providing a growing selection of items, *The Whole Earth Catalog* became a tool in itself, enabling the collective and the individual to evolve in any possible direction and on the way create new tools, essentially fulfilling Brand's philosophy "that truly adaptive systems grow from the bottom up, not from the top down" (Brand 1994, p. 5).

Steve Jobs saw the catalog as one of the main sources of inspiration of his time and describes it as a "sort of […] Google in paperback form, 35 years before Google came along" (Stanford Report 2005). Kevin Kelly, the founding executive editor of *Wired* magazine, makes similar comparisons, calling it a "user-generated website before the web, before computers, employing only cheap newsprint" (Kelly 2010, p. 3). The impact that the publication had on people of the time, who dropped out of society to enjoy low-tech, back to the roots lives, only to return ten to twenty years later and become Silicon Valley entrepreneurs, is very compelling especially from a cultural point of view. In the foreword to the to date latest issue, *The Millennium Whole Earth Catalog*, published in 1994, Brand himself draws direct analogies in between the experimental spirit of the communes "who were setting out to reinvent civilization" and the rise of the personal computer revolution that wanted to "undermine the high priests […] of information technology and hand their power to absolutely everybody" (Brand 1994, p. 5). In that sense *The Whole Earth Catalog* synthesized a vision of technology, and especially the computer, as a counter-cultural power, democratizing information and empowering its users to establish and maintain their independence from large institutionalized systems. It was an outlet, a common ground, a means of expression and portrayal of the spirit of a society and time that was eagerly searching for alternatives on all fronts (Scott 2008, p. 166).

However, especially in architecture this desire for reformation and change could already be felt much earlier. It was at the **CIAM** IX meeting in Aix-en-Provence in 1953 when the ideas of a young generation of architects, many of whom would later become core members of **Team 10**, clashed prominently with key figures of the modernist movement, such as Le Corbusier, Sigfried Giedion, and Walter Gropius (Frampton 2007, p. 271). Three years later at CIAM X in Dubrovnik, Yona Friedman proposed an architectural model that would be capable to respond to social, occupational, infrastructural, or other related changes more flexibly than existing tendencies. His vision of the 'Ville Spatiale,' the spatial city, is centered on the idea that architecture should solely provide a framework for self-expression and self-fulfillment and liberate its inhabitants from master planning and spatial paternalism. The three-dimensional structure is raised from the ground and through this additional dimensionality expands the area of cities and provides different levels of usage and occupation, allowing for example the superposition of an industrial zone above a residential or commercial area. The continuous, homogeneous, and endlessly extendable system thus not only provides great flexibility and versatility but also authorizes the unrestricted growth of the urban settlement. Friedman insists in his approach, like Kurt Schwitters before him, on the importance of the process rather than the final result (Friedman 2006, p. 23). Further stressing his affinity to the work of Schwitters, he later coins the term 'Merzstrukturen,' merging the, in his eyes, two most significant impulses to twentieth century architecture, the benefits of industrially fabricated space-frame structures, popularized by Konrad Wachsmann, and the extreme emotional individualism of Schwitters' Merzbau. Friedman despises any kind of generalized anthropomorphic system, such as Le Corbusier's Modulor or Da Vinci's Vitruvian Man, and strongly believes in the uniqueness of the individual, whose interest and future behavior cannot be predicted (Wigley 1998, p. 41). His concept, which he further elaborates in his *manifesto on mobile architecture*, thus describes a solid but flexible structural system within which the tenants are encouraged to create their own environments, all different but highly personalized. Like his Dutch colleague Constant Nieuwenhuys, Friedman trusts in the progress of automated production and the resulting increase in leisure time, which thus should mark the core theme of any future society. Moreover he argues that new metropolitan cities have to be free from the influence of urban planners, include designated agricultural areas, be climatically controlled, comply to a clearly defined grid and scale, rise above an existing city, and aim for an average population of three million people (Friedman 2006, pp. 65–69). Friedman believes that such guidelines, while intended to remain open and encourage improvisation, are necessary to avoid chaos and disorder. To describe these frameworks, which go beyond classical architectural means of expression, like plans or drawings, he proposes the use of cooking instructions, recipes, or tutorials that allow for interpretation and experimentation. In order to prove his vision in a real world context he establishes simple architectural systems for developing countries, using locally available materials and traditional building techniques, and visualizes them through sets of self-explanatory comics.

Friedman's aversion toward a top–down dictatorial design, of architects superimposing their ideas of form, function, and aesthetics over the immanent needs and desires of constantly evolving individuals has many parallels to the philosophy of Bernard Rudofsky. In *Architecture Without Architects: A Short Introduction to Non-pedigreed Architecture* Rudofsky reveals the functional and artistic richness of 'primitive' dwelling, as an attempt "to free ourselves from our narrow world of official and commercial architecture." The book, which stems from an eponymous exhibition at the Museum of Modern Art, New York in 1964, graphically presents a wide array of vernacular styles from all around the globe, including Peruvian amphitheaters, Swedish cemeteries, or inhabitable rock formations in Spain. Rudofsky argues that even though vernacular buildings are generally derided as being fairly primitive their closer examination provides invaluable insights, which have largely been lost within the modern discipline, like for example an intuitive talent of carefully embedding buildings within their natural environment instead of conquering and dominating nature. Rudofsky disputes further that since vernacular architecture doesn't need to respond to fashion or trends "it is nearly immutable, indeed, unimprovable, since it serves its purpose to perfection." Most of all he, however, emphasizes the "humaneness of this architecture," which has evolved and improved throughout hundreds of generations and is not only aware of its influence on itself and its surroundings but also its limitations and boundaries. Concluding he declares that what can be learned from such structures goes far beyond economic and aesthetic aspects but involves an inherent wisdom on how to live and mutually prosper within a communal society (Rudofsky 1964, pp. 1, 4, 8). Even though *Architecture without Architects* is much less of a practical handbook than other publications of the time[2] and possibly promotes and idealizes certain building types without much reflection on their particular context and current relevance, it does provoke the reconsideration of existing structures and encourages a refocus on aspects like humanity and empathy.

While the approaches of Friedman, Brand, and Rudofsky vary in their orientation and focus, they all share a common interest in revealing knowledge and using it to promote the emancipation from patronizing or controlling structures. In Brand's case *The Whole Earth Catalog* is a fully curated conglomeration of related artifacts supplementing and enhancing one another, Rudofsky reveals particular aspects of lost or forgotten skills and techniques through a broad collection of photographs and brief explanatory excerpts, and Friedman develops simple, illustrated 'how-to' manuals in the form of pamphlets and comics.

3.3.1 Approaches for the Mediation of Material Knowledge

Today, in relation to access to materials, common tendencies for the conveyance of material knowledge include libraries, catalogs, and databases, both online and offline. Within these, materials are usually sorted and categorized according to particular similarities. In printed form this is often based on either their dominant material class or other defining physical properties. In respect to traditional materials such types of categorization, according to certain meaningful properties, and a two-dimensional representation in the form of photographs and technical drawings work quite well. Particularly when coming from a design or architecture background, one can often build upon physical experience and anticipate how a material feels or behaves simply by looking at it or reading about its properties. Regarding *information materials* however, which are far less known and available, it is much more difficult to foresee their usage, especially since they are usually displayed among static materials, employing the same kind of categorization. This not only leads to a severe generalization for the sake of comparability but also encourages people to understand them like static materials and thus neglect their behavioral aspects. In this context it is hence not only necessary to grant access to the relevant information but also provide a suitable framework to navigate and compare. For this framework to be understandable it should be based upon familiar principles or structures while representing a distinct orientation and unique identity in order to promote thinking in terms of dynamic processes instead of static properties.

[2]Steve Baer's *Dome Cookbook* (1967), which was reviewed in the first *Whole Earth Catalogue*, provided comprehensive, hand drawn instructions for the creation of a variety of domes in different shapes, sizes, and forms. Ant Farms' *Inflatocookbook* (1971) is a collection of the group's experience in designing, building, and erecting inflatable structures for different contexts. William Powell's *The Anarchist Cookbook* (1971) is a manual that contains information on the DIY manufacture of explosives, drugs, and telecommunications devices. It was written as a protest against the US army involvement in the Vietnam War.

3.3.2 The Materiability Research Network: Access to Information Materials

The *materiability research network* (http://www.materiability.com) is a community platform, an educational network, and an open materials database that provides in-depth access to emerging material developments. The network was instigated in 2012 with the prime demand to provide unrestricted access to knowledge concerning *information materials* and to form the base for a community that shares a fascination with dynamic material phenomena, digital creation, and physical making. Since the materials generally come from different areas of research, bridging chemistry, materials science, biology, engineering, or robotics, the aim of the network is not in clearly identifying and categorizing them as such but rather in demystifying and revealing their complexity while fostering interdisciplinary exchange and cooperation. The website forms a continuously growing database on a wide range of materials, provides hands-on tutorials to self-produce them, and promotes their assembly in speculative spatial installations and artistic explorations. This trinity, inspiration, information, and instructions is ought to help develop a common language in order to communicate across disciplines and bridge the gap between research, education, and practice. The website is divided in a publicly accessible part and a members-only area, as Fig. 3.2 displays.

Registration to the network, which is free of charge, allows members to unconditionally access all provided information, exchange with fellow members via an internal messaging service, engage in forum discussions, and also submit their own related projects or experiments. The core content is split into the three categories 'projects,' 'materials,' and 'tutorials.'

'Projects' are openly accessible and focus on innovative and inspirational usage of *information materials* in an architectural or design context. They can be sorted according to rating (in a range from one to five), views, and age, as well as several tags in eight main categories, which are displayed in the top-right corner of each post:

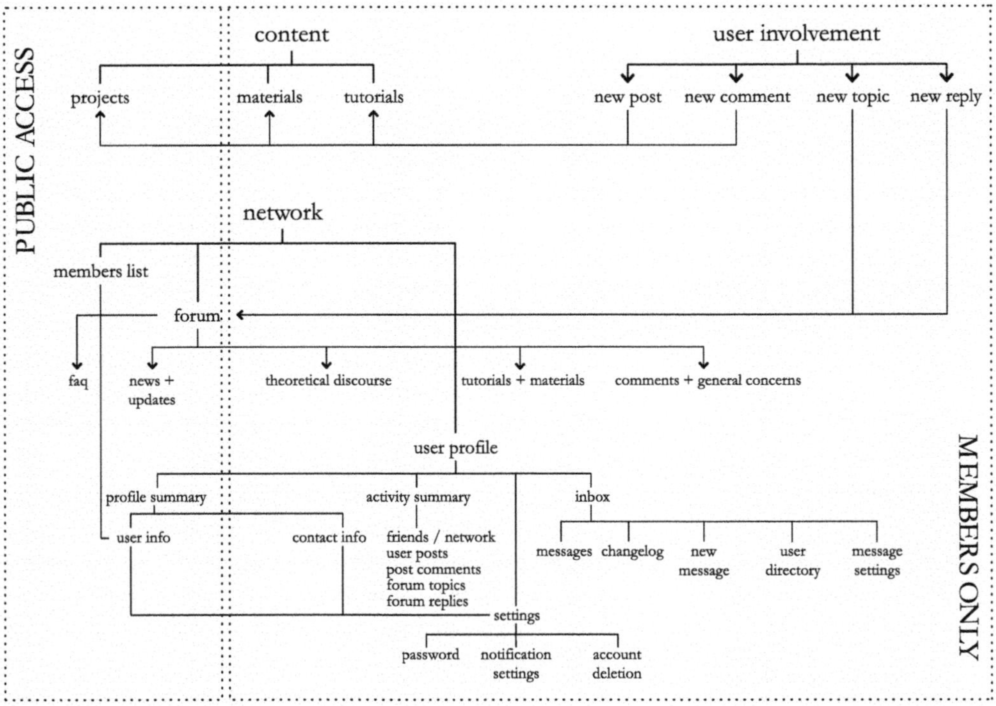

Fig. 3.2 Website structure of the materiability research network, displaying content hierarchy, and different user access areas

1. Material Class (ceramic, metal, organic, polymer, semiconductor, textile)
2. Creative Discipline (architecture, art, design, engineering, science)
3. Phenomenological Behavior (adhesion-changing, color-changing, electricity-generating, energy-exchanging, light-emitting, matter-exchanging, shape-changing, sound-creating, static)
4. Physical Property (flexible, fluid, soft, rigid)
5. Optical Property (opaque, translucent, trans parent)
6. Scale (small, medium, large, x-large)
7. Status (realized, conceptual, information)
8. Institution

Once a specific tag has been picked all related projects are displayed, but it is also possible to choose multiple tags and show projects that posses either all or any of the selected descriptions. 'Materials' and 'tutorials' are only visible to active members of the network, who are registered and logged-in. This has several reasons:

- First the provided information is intended mainly for educational and noncommercial purposes and thus targets students or otherwise academically affiliated people.
- Second requiring users to register fosters the establishment of a distinct network and associated allegiance. The website thus emancipates itself from being yet another material database but turns into a social community with its own identity.
- And finally disclosing particular content from the public decreases the website's vulnerability, enables internal exchange, and promotes the distribution of individualized content to selected parties.

The information on materials is intended to present a scientifically grounded context, providing in-depth but easily understandable details. The graphically illustrated essays, which are essentially trimmed versions of the material chapters within this book, are usually divided into a brief material introduction, an excerpt on its historical development and usage, an explanation of its structure and operational principle, a short description of its fabrication, and an excursion into existing and potential (architectural) applications. In addition to this theoretical material information, related tutorials are offered, which are often developed together or with the help of experts from the respective fields. Each tutorial follows a similar setup and involves both photographic and text-based step-by-step instructions.

3.4 Information Material Literacy

> Science and technology multiply around us. To an increasing extent they dictate the languages in which we speak and think. Either we use those languages, or we remain mute.
>
> Ballard (1974, Crash, p. 7)

Literacy is classically described as the ability of a person to read and write a short, simple statement about one's everyday life with understanding of its meaning (United Nations 2008). Since the 1950s the global adult literacy rate, which is referring to people aged fifteen and above, possessing such basic reading and writing abilities, has increased from 56 % over 70 % in 1980 (UNESCO 2005) to 84 % in 2011 and is exceeded 86 % in 2015. For comparison, the global youth literacy rate, which is only counting the population aged fifteen to twenty-four years, has grown from 83 % in 1990 to 90 % in 2011 and 92 % in 2015 (Huebler and Lu 2013, pp. 10, 18–19). Even though this means that as of 2011 around 774 million people were still lacking fundamental literacy skills, which is roughly equivalent to the whole population of Europe, the UNESCO is confident that with older generations making place for younger, more educated people, worldwide literacy will continue to increase and prosper. The UNESCO, which has since it's founding in 1946 played a key role in progressing adult learning, understands literacy as a basic human right, as a tool of individual empowerment, and as the core of personal and social development. To them the creation of "literate environments and societies is essential for achieving the goals of eradicating poverty, reducing child mortality, curbing population growth, achieving gender equality and ensuring sustainable development, peace and democracy" (UNESCO 2005).

A person's ability to read and write thus has much farther-reaching implications than the basic skill itself implies. Literacy is like a door that, when unlocked, reveals previously unknown and inaccessible territories, exposing a richness of new possibilities and choices. If one was to consider language as the core technological advancement of humanity, then literacy can be seen as a learnable competency that facilitates the meaningful expression of language through the usage of a tool (like a pen, typewriter, or computer), which can be understood and taken further by other literate people. In that respect any kind of technology can be regarded as being dependent on users who are literate in its application to fully exhaust its potentials. The more versatile a technology is, the greater are the freedom and opportunities it offers, yet also the more complex are its usage and associated literacy.

In *What Technology Wants* Kevin Kelly refers to the Internet as the major new technological development of the twenty-first century, which most of all provides new possibilities, empowering people to connect, share, and collectively progress. People use the Internet not because they have to or because there are no other options but because they want to. It is a conscious decision to emerge within its digital sphere and nurture from the seemingly limitless freedom it offers. Kelly makes a similar observation in the phenomenon of people moving to cities and urban areas. For many people, who migrate to cities and end up in slums or informal settlements, the living conditions turn out to be much worse than in the villages or towns they come from. Yet they accept these sacrifices due to the much greater amount of options and choices urban environments offer, not necessarily for themselves, but especially for their offsprings (Kelly 2010, pp. 80–83).

Modern metropolises are extremely productive technological melting pots which not only generate the bulk of the global economic and innovative output but also form highly connected environments where the boundaries between physical (offline) and digital (online) world are continuously blurring. John Palfrey and Urs Gasser, authors of *Born Digital: Understanding the First Generation of Digital Natives*, emphasize the necessity for young people to learn what they call a 'digital literacy' in order to be able to thrive within these new high-tech settings. To them it is not sufficient to have access to a certain technology but to know how to employ it purposefully. This involves experience in navigating the digital landscape, knowledge in processing and abstracting the provided information, and if necessary knowing how and where to dig deeper. More importantly, however, are an awareness of one's actions and implications, including issues of privacy, safety, exposure, or surveillance. This anticipation of potential consequences might be one of the major challenges when learning digital literacy not only because predictions are hard to make but especially since the social core of the Web is dependent on user-generated content. The Internet is thus much more than a place that provides quick access to an endless stream of information, it marks a vibrant environment that fosters creative learning, self-expression, personal autonomy, collaborative engagement, and cultural exchange. By empowering its users to become creatively involved in the process of its creation it is shifting "us away from a world of largely passive consumers of content produced by a few powerful professionals toward communities of increasingly active users" as Palfrey and Gasser observe (Palfrey and Gasser 2008, pp. 15, 114).

3.4.1 Collective Making in the Third Industrial Revolution

Long-time *Wired Magazine* editor in chief, Chris Anderson elaborates upon the productive potential of the digital world and its effects on society in his book *Makers*. He claims that through the emergence of consumer focused digital fabrication tools, such as desktop 3D printers, as well as instant access to cloud-based manufacturing stores, information that was formerly constraint to a largely digital existence now finds its way back to the physical world. To him this transformation, from bits back to atoms, forms the logical extensions of the digital realm and will soon replace mass production through mass customization, turning into, what he calls, the

'Third Industrial Revolution.' Comparing this particular development to ground-breaking historical events like the invention of the Apple II or the birth of the laser printer and its effects on the publishing industry, he predicts it will convince consumers to move away from cheap mass produced products toward niche designs and the work of artisans (Anderson 2012, pp. 17–18).

What becomes clear is that strategies that are successful in the digital world, such as networking, exchange and open ownership, can be adapted for the creation of physical artifacts, which is especially interesting in respect to architectural design education. Through the access to new tools, community-based software and scripting languages, as well as CNC and rapid prototyping machines we can experience a new relationship between the digital and the physical, which involves a reconsideration of representation and manifestation or drawing and making. Yet what sometimes tends to be missing in these explorations is a thorough examination of materiality, which with the prevalent tools that process sheets or print powdered granulates, becomes increasingly difficult. Even though the idea of collective sharing and development is certainly something architects should deeply identify themselves with, the praise of digital machines, and in particular 3D printing techniques, bears some questions, if not doubts, especially when it comes to leaving the representational scale of model making. But it is exactly this threshold, the scale in between models and real buildings, the scale of temporary installations, pavilions, and sculptures that provides an enabling opportunity to experiment, express, and experience architecture. It is through these 1:1 prototypes that designers can verify whether their abstract ideas can be realized and through hands-on experience they can draw conclusions on how to improve the operation of tools and particularly the application of materials. So just as the Internet requires users who are sensitive and conscious of their actions and the environment they are engaged in, any kind of technology should be evaluated in a larger, more comprehensive context (Kretzer 2013, p. 120).

The true power of the Third Industrial Revolution might thus be less of economic importance, as Anderson claims, but in forming the reflective base for collective physical exchange through digital means, allowing people to seamlessly swap and modify their design data without constraints in time and space. People are thus able to work together, support and criticize each other, and build upon one another's expertise much more fluently and directly than ever before. For such tendencies to become more than just a shared joy in making but to turn into actual business models will, however, require fundamental reconsiderations of established structures, such as hierarchy, authorship, or liability. Yet for the time being and especially in the context of the present chapter the focus shall be on their intellectual and social empowerment, rather than their economical or political effects. It is the increase in choices and possibilities, the awareness of like-minded others, the location- and time-independent availability, and the reassuring knowledge of one's own unlimited ability due to the above, that are truly revolutionary.

3.4.2 The Materiability Research Network: A Growing Community

The *materiability research network* is built upon the above-described phenomena and directly draws from their success. The platform, which is based on a slightly modified WordPress framework and a number of customized plugins, has since its instigation on June 13, 2012 grown steadily to 1700 registered members as of October 2016, as shown in Fig. 3.3.

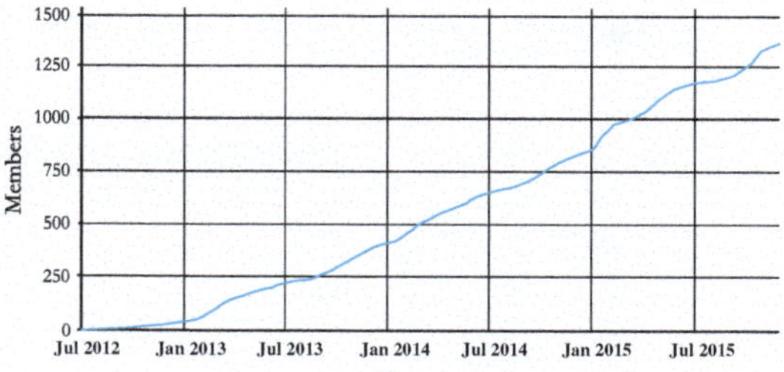

Fig. 3.3 Accumulative growth of the materiability research network based on membership registrations from June 2012 until mid-November 2015

Visitors of the website are highly international, as shown in Fig. 3.4. Due to this and the networks other associated social channels a large amount of people can be reached and updates can be disseminated vast and quick. To promote exchange and collaboration among the community, the network provides an internal messaging service, allowing members to directly get in touch with each other.

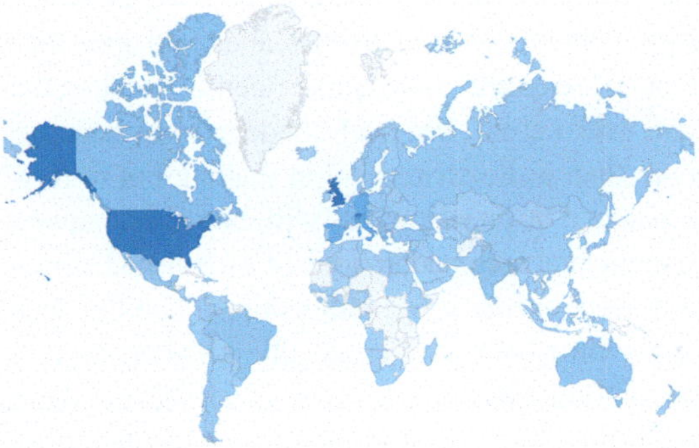

Fig. 3.4 Map overview of website visitors in the period between June 13, 2012—May 12, 2015 with darker areas representing higher frequency

The *materiability research network* thus not only represents a website that provides information on a specific topic, but it encourages its members to get actively involved in its development. Becoming a member of the network means to identify oneself with the content and community. By publishing information on the platform users have to take responsibility for their work and review and expose it within the larger context (Fig. 3.5). This means that they have to be aware of the information they consume but also the content they produce and its

Fig. 3.5 The website's front page as of February 2016 provides an overview of the most recent submissions in the categories 'projects,' 'tutorials,' and 'materials' as well as a feed displaying new forum topics

further impact. The network provides a steadily growing overview of *information materials* in an architecture and design context in the form of tutorials, information, and projects, while encouraging its members to exchange and critically reflect upon their potential usage. This is ought to provide a substantial basis for an *information material literacy.*

3.5 Summary and Conclusion

> Rule six: Nothing is a mistake. There's no win and no fail, there's only make.
>
> Kent and Steward (1992, Learning by Heart, p. 176)

Just as the environment is constantly changing and evolving, architecture and especially architectural education needs to progress, too. Particularly the mediation of technological advancements and the ever-increasing amount of materials require new concepts and methods for teaching and the dissemination of knowledge. Since many technological developments take place in areas that are not classically linked to architecture, such as materials science, biology, or chemistry, cross-disciplinary exchange and collaboration are key qualities that need to be communicated. Successful exchange and interaction require most of all shared interests, similar background knowledge, and the ability to converse in a mutually understandable language. The present chapter proposes the conveyance of such proficiency, particularly in respect to *information materials*, based on three interrelated concepts that in their unity are referred to as the *materiability* approach.

- The first strategy highlights the importance of a process-oriented model for physical making and hands-on experimentation. Focusing on continuous exploration rather than a final product, allows students to understand and emphasize dynamic material properties and develop (adaptive) architectural concepts in regards to time and space.
- The second aspect emphasizes the need for open and flexible access to information in order to democratize and decentralize knowledge. This not only promotes collective exchange and identification with a common cause but also encourages the liberation from established structures and the development of non-authoritarian alternatives.
- The final part elaborates upon the relevance of being literate in the usage of *information materials*, which exposes a myriad of new possibilities and choices for their usage. The demonstrated *information material literacy* constitutes the result of the preceding two approaches, the physical experience one gains when dealing with *information materials* and the unlimited access to similarly comprehensive information on further materials.

The concluding assumption is that due to such literacy, it becomes possible to abstract and anticipate the functionality, behavior, and usage of any type of *information material* without necessarily requiring a working prototype, just as one, with a little experience in craftsmanship, does not require to touch a piece of wood, stone, plastic, or fabric in order to know how it feels, looks or what it can be used for. This idea correlates with Kevin Kelly's reasoning on growing urbanization. Obviously people move to metropolitan areas due to an increase in possibilities and choices, yet they most likely will not ever be able to fully exhaust all that is available (Kelly 2010, pp. 80–83). The decision to reside within an urban setting is thus less related to the actual process of exploiting its resources than the conscious awareness of the possibility. The knowledge of potentially being able to use anything that is on hand, is the real freedom and luxury such environments offer. The Internet represents a similarly vital place, empowering its literate users to find and access whatever they are interested in, independent from time and location.

Once enough information, data, and experience on the usage of *information materials* is made accessible it will also become possible to predict their purpose in a much more progressive manner. This will eventually free them from persistent associations to mechanically infused paradigms, which to date still mark the main form of describing dynamics, movement, and behavior. The *materiability* approach as a didactic model has been tested and verified throughout numerous workshops and student courses, which are displayed in detail in the next chapter. Despite its brevity of a little more than five years, it can so far be considered a largely successful method to engage students in exploring new territories but obviously always needs to be adapted to the respective context and situation. Especially the display of dynamic behaviors requires further investigation and will need the development and acceptance of alternative means to the classical architectural plan and section display. One possibility would be to shift the focus toward working model demonstrators and physical prototypes instead of their graphical representation. Another option is to draw inspiration from the Internet and online representations, which allow the seamless combination of video, audio, image, and text-based content, an approach that is strongly pursued by the *materiability research network*. The main purposes of the platform, the dissemination and access to knowledge, the cross-linkage of like-minded people, and the establishment of a shared identity are all aspects that are already successfully and continuously progressing. In that respect the *materiability research network* has found and conquered its niche in which it will continue to prosper and hopefully inspire.

References

Anderson C (2012) Makers: the new industrial revolution. Crown Business, New York

Ballard JG (1974) French ddition: Crash (trans: Ballard) (1995). Vintage, London

Ballard Bell V, Rand P (2014) Materials for design, vol 2. Princeton Architectural Press, New York

Bonnemaison S, Eisenbach R (2009) Installations by architects. Princeton Architectural Press, New York

Brand S (1968) Whole earth catalog: access to tools. Whole Earth

Brand S (1994) Civilization and its contents. In: Howard R (ed) The millennium whole earth catalog, vol 5. Whole Earth

Burns Gamard E (2000) Kurt Schwitters' Merzbau: the cathedral of erotic misery. Princeton Architectural Press, New York

Clear N (2014) Convergence: architecture as integrated spatial design. In: Spiller Neil, Clear N (eds) Educating architects: how tomorrow's practitioners will learn today. Thames and Hudson Ltd., London, pp 92–101

De Zegher C, Butler C (2010) On line: drawing through the twentieth century. The Museum of Modern Art, New York

Elger D (2004) Dadaism. Taschen, Cologne

Fisher T (2000) In the scheme of things: alternative thinking on the practice of architecture. University of Minnesota Press, Minneapolis

Frampton K (2007) Modern architecture: a critical history, 4th edn. Thames and Hudson, London

Friedman Y (2006) Pro domo. Actar, Barcelona

Hollein M (2002) In: Lynn G, Rashid H (ed) Architectural Laboratories. NAI, Rotterdam, p 7

Huebler F, Lu W (2013) Adult and youth literacy: national, regional and global trends, 1985-2015. UNESCO Institute for Statistics, Montreal

Internet World Stats, Usage and Population Statistics (2012) World internet usage and population statistics. http://www.internetworldstats.com/stats.htm. Accessed 10 Mar 2014

Kelly K (2010) What technology wants. Viking, New York

Kent C, Steward J (1992) Learning by heart: teachings to free the creative spirit. Bantam, New York, 205pp

Kretzer M (2013) Open matter(s): a community approach towards smart materials and their potential for architecture and design. In: Stacey M (ed) Prototyping architecture: the conference papers. Building Centre Trust, London, pp 118–130

Kretzer M (2014) Beyond performance. In: Kretzer M, Hovestadt L (eds) ALIVE: advancements in adaptive architecture. Birkhäuser, Basel, pp 72–77

Lynn G, Rashid H (2002) Architectural laboratories. NAi, Rotterdam

Moser C (2014) Architecture 3.0: the disruptive design practice handbook. Routledge, New York

Ochsner JK (2012) Furniture studio: materials, craft, and architecture. University of Washington Press, Seattle

Palfrey J, Gasser U (2008) Born digital: understanding the first generation of digital Natives. Basic Books, Philadelphia

Porritt J (2013) The world we made: Alex McKay's story from 2050. Phaidon Press, London

Rudofsky B (1964) Architecture without architects: an introduction to nonpedigreed architecture. Museum of Modern Art, New York

Scott FDE (2008) Living archive 7: Ant Farm—allegorical time warp: the media fallout of July 21, 1969. Actar, Barcelona

Schwitters W (1924) English edition: El Lissitzky: life, letters, texts (trans: Lissitzky-Küppers S) (1968). New York Graphic Society, Greenwich, p 351

Stanford Report (2005) You've got to find what you love. Jobs says. http://news.stanford.edu/news/2005/june15/jobs-061505.html. Accessed 20 Mar 2015

The Network: Cisco's Technology News Site (2012) Cisco's VNI forecast projects the internet will be four times as large in four years. http://newsroom.cisco.com/press-release-content;jsessionid=4DFB3651CC7386369DFB9A81B1815B74?articleId=888280&type=webcontent. Accessed 10 Mar 2014

Turner F (2006) From counterculture to cyberculture: Stewart Brand, the whole earth network, and the rise of digital utopianism. University of Chicago Press, Chicago

Umemoto N, Reiser J (2014) Some thoughts on pedagogy. In: Spiller N, Clear N (eds) Educating architects: how tomorrow's practitioners will learn today. Thames and Hudson Ltd., London, pp 212–219

UNESCO (2005) Education for all—EFA global monitoring report 2006: literacy for life. UNESCO Publishing, Paris

United Nations (2008) Principles and recommendations for population and housing censuses: revision 2. United Nations, New York

Wigley M (1998) Constant's New Babylon: the hyper-architecture of desire. 010 Publishers, Rotterdam

4 A Selection of Emerging Information Materials, Their Properties, Fabrication, and Application in Speculative Spatial Installations

4.1 Introduction

> In any creative medium, there has to be substance for the work to hold up. For example, a piece of music is not an answer. It's a stimulus. It leads you to thinking and feeling a certain way, which you wouldn't have done without that particular experience.
>
> Woods and Flom (2004), The Reality of Experimental Architecture

The following is an attempt to instigate a more comprehensive understanding of *information materials*, revealing commonalities that fundamentally distinguish them from the classical notion of materiality, even though they might come from very different areas and exhibit formally unrelated properties. By providing a broad spectrum of materials, yet describing each of them fairly extensively in both theoretical and practical terms the reader is encouraged to discover similarities and develop an independent view and experience on their functionality and potential usage. As of now the collection includes eight materials, which could, if categorization was important, be sorted into five phenomenological behaviors: kinetic movement (*Electroactive Polymers and Soft Robotics*), light emittance (*Bioluminescence and Electroluminescent Displays*), color change (*Thermochromics*), electricity creation (*Dye-sensitized Solar Cells*), and structural performance (*Aerogels and Bioplastics*). Each material description is organized into a brief material history, an explanation of the material's structure and operation principle, its fabrication, including step-by-step instructions on how to self-produce the material, as well as an overview of current and potential applications in architecture and design. The usage of some materials is further exemplified through a number of case studies, such as workshops or student projects.

The selection of materials provides a broad range and is based on the interest into their specific phenomena and possible accumulative combination due to shared properties or appearance, like being film-based or responding to electrical inputs. Most of the materials were utilized during workshops and courses as described in Sect. 3.2, *The process of physical making*. Since the goal of these courses is the mediation of the particular properties of *information materials* and their potential in the context of adaptive architectural systems through hands-on experience it was important that the materials could be fabricated following rather simple rules without the need for extensive technical equipment. Hence, the majority of the following tutorials can be executed in informal, temporary setups. The focus of the provided instructions is thus neither on improving a material's properties in relation to efficiency or durability nor on developing new materials for specific purposes but solely in providing direct experience of a material by breaking down scientific processes into easily understandable steps. The present information can be used as a primer for teaching the working principles of

M. Kretzer, *Information Materials*, DOI 10.1007/978-3-319-35150-6_4

information materials in an applied educational context similar to the courses described in the respective *Case Studies* sections.

The hope is that the following sections not only provide interesting insights into various emerging material technologies but especially initiate a change in thinking about *information materials* and pave the path for an *information material literacy*.

4.2 Aerogels

Aerogel (Fig. 4.1) is an ultralight material, synthetically produced from a gel in which the liquid component has been removed without collapsing its structure. The transparent, porous material weighs with 3 mg/cm^3 only three times more than air and is hence among the lightest solids ever created (Han 2013, p. 703), only ousted in 2012 by the even lighter aerographite with 0.18 mg/cm^3 (Mecklenburg et al. 2012, p. 3487), and in 2013 by graphene aerogel with 0.16 mg/cm^3 (Sun et al. 2013, p. 2554). It exhibits outstanding low density and is generally made of amorphous silicon dioxide, and 99.8 % air, yet it is remarkably strong structurally, and can sustain thousands of times its own weight. In theory, a piece weighing less than 500 g could withstand 500 kg in pressure (American Chemical Society 2012). Since it negates basically any kind of energy transfer, whether being thermal, electrical, or acoustic, a 6-mm sheet would provide as much insulation as 8 cm of fiberglass (Dodson 2012). Based on its particular microstructure, aerogel is a strong desiccant, rapidly absorbing any moisture in its environment. Its melting point is at 1200 °C and it comes as monoliths, granulates, or powders. Aerogel feels like a very light, hard Styrofoam. Until 2011, the material held 15 entries in the *Guinness Book of World Records*.

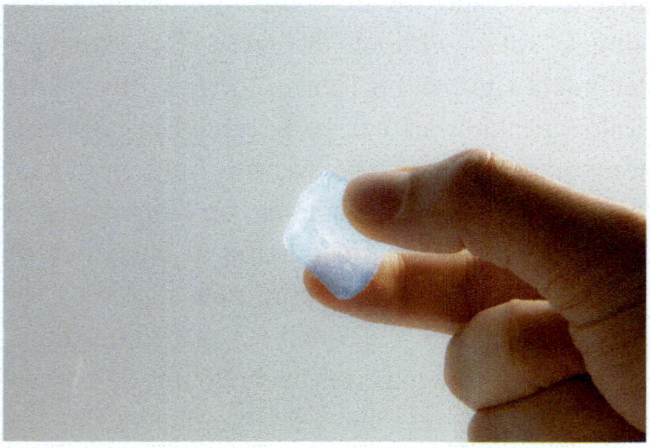

Fig. 4.1 Aerogel is ultralight and appears *blueish* due to its structural arrangement, which scatters the shorter wavelengths of visible light (Kretzer 2013a, b)

4.2.1 History

The term aerogel was first introduced by the American scientist and chemical engineer Samuel Steven Kistler (1900–1975) in 1931, in an article published in *Nature* (Kistler 1931, p. 741). Kistler, who at that time taught chemistry at the University of Illinois, used the word to describe a gel in which the liquid part had been replaced by a gas, without affecting the gel's microstructure. Up until then, wet gels had been dried by evaporation, but Kistler used a new supercritical drying technique, in which the liquid that permeated the gel was vacated after being converted to a supercritical fluid (Pierre 2011, p. 3).

Besides producing transparent silica aerogel with densities as low as 30 mg/cm^3, Kistler's research included the creation of alumina aerogel, tungsten oxide aerogel, iron oxide aerogel, and tin oxide aerogel, as well as aerogel from organic components like cellulose, nitrocellulose, gelatin, agar, or egg albumin (Pierre 2011, p. 5).

In 1935, Kistler left his teaching position and joined the Monsanto cooperation in the early 1940s where he began developing granular silica aerogel. The commercialization of aerogel began in 1942 with Monsanto producing Santocel. Monsanto described the product as a very effective heat insulating material but also used it as flatting agent in paints and varnishes or as additive in toothpaste and cosmetics or even napalm. However due to the high costs in production and the emergence of much more inexpensive silica, the company discontinued its research into aerogel at the end of the 1960s.

Interest in the material, and especially its very low thermal conductivity, reemerged in the late

1970s as energy conservation became more and more important. One of the major breakthroughs in aerogel research was the application of sol-gel chemistry to silica aerogel preparation by Stanislaus Teichner and G.A. Nicoloan in 1972, which allowed its production in one continuous process (Han 2013, pp. 701–702). These techniques were further refined and made safer during the 1980s, creating aerogel 30 times lighter than before and developing new types of organic and carbon aerogel with extraordinary new properties. Motivated by such promising developments, new ideas in commercializing aerogel were born, along with scientific awareness from a growing body of researchers fascinated by the material's unique features and the range of substances that could be used for its production. In 1985, the first *International Symposium on Aerogels* (*ISA*) was organized by Professor Jochen Fricke and held in Würzburg, Germany. It attracted 25 international researchers to present their work (Aerogel.org).

The 1990s were the decade that aerogel became publicly popular. In 1991, Dr. Tillotson at Lawrence Livermore National Laboratory developed the process for synthesizing ultralow density silica aerogel, which, with just 3.1 mg/cm^3, for a long time held the Guinness World Record for lowest density solid (Tillotson and Hrubesh 1992, p. 44). In 1996, a rapid supercritical extraction technique was developed, speeding up aerogel production to only 4 h in time. 1-year later, in 1997, NASA successfully employed aerogel insulation to shield the electronics of the Sojourner rover and two more vehicles in the following years as part of their Mars Pathfinder mission. In 1999, NASA used aerogel for the comet dust collector of the Stardust mission, which reentered Earth's atmosphere in 2006 bringing along various stellar particles, embedded in its aerogel tiles (Burchell et al. 2006, p. 390).

By crosslinking silica aerogel with polymers Prof. Nicholas Leventis at NASA's Glenn Research Center developed x-aerogel in 2002, a new class of flexible, mechanically stable, porous aerogel that can be made into a film. X-aerogel is robust against folding, crushing, or creasing and about 500 times stronger than silica-based aerogels (NASA Glenn Research Center). In 2003, Prof. Stephanie Brock at Wayne State University in Detroit prepared photoluminescent, semiconducting aerogel from metal chalcogenide, which has potential as chemical sensors. In 2007, researchers at the University of Pennsylvania assembled carbon nanotubes into a gel structure and then supercritically dried them to prepare carbon nanotube aerogel. In 2009, Prof. Nicholas Leventis announced the creation of the first metal aerogel made from iron (Han 2013, p. 705). Since 2010, the Swiss Federal Laboratories for Material Testing (EMPA) together with the Swiss render manufacturer Fixit AG are developing a spray-on aerogel-based plaster that provides two to three times better insulating properties than conventional materials on the market (EMPA Materials Science and Technology). Early 2013, researchers at the Department of Polymer Science and Engineering, Zhejiang University, China reported the creation of ultra-flyweight aerogels by freeze-drying aqueous solutions of carbon nanotubes and graphene oxide sheets (Sun et al. 2013, p. 2554).

Further research into aerogel currently includes its practical use for the containment of nuclear waste, CO_2 trapping, water-repellent coatings, chemical sensors, metal casting molds, acoustic transducers, energy storage devices, or pharmaceutical drug carriers (Pierre 2011, p. 12). Aerogel is being commercialized by a number of companies, like the Cabot Corporation, which produces aerogel as insulators for windows, Nano HiTech, and Okagel, which also make insulation products, the American Aerogel Corporation that makes open-cell foam materials, the Birdair Company, producing membranes containing aerogel inserts, and the Aspen Aerogels Company, which specialized in flexible insulation products.

4.2.2 Material Structure and Synthesis

Aerogel's unique properties, like outstandingly low thermal conductivity and sound velocity while being largely transparent and extremely lightweight, result from the material's dendritic microstructure, in which spherical particles are joined into clusters that form three-dimensional, utterly porous fractal chains, too small for air to travel through. This particular structural arrangement, which breaks the shorter wavelengths of visible light and gives the gel its smoky blue color, however also makes aerogel very fragile and prone to shattering. The most common types of aerogel are based on silica, carbon, and metal oxides, with silica aerogel being the most extensively studied and applied. Carbon-based ones are black and exhibit electrically conductive capabilities together with an extremely high surface area, making them useful for supercapacitors or fuel cells. Metal oxide aerogels can be magnetic and depending on the dopant have differently bright colors.

The structural network and thus the size and density of the pores, is formed by the condensation of primary particles at nanometer scale and controlled by a chemical process, usually the sol-gel process.

4.2.2.1 Sol-Gel Process

The main advantage of the modern sol-gel process (Fig. 4.2), compared to Kistler's original procedure, is that it skips the solvent exchange step, which drastically speeds up the entire method.

The first phase of the procedure involves the colloidal dispersion of solid nanoparticles in a liquid solution (1). In the case of silica aerogels, an alcohol is mixed with a silicon alkoxide precursor. Through a hydrolysis reaction (2) and consequent condensation (4) the sol (3) then gradually transforms into a gel-like system (5), an alcogel (or aquagel/hydrogel if other liquids are used).

The earliest precursor, applied by Kistler, was sodium metasilicate (Na_2SiO_3), also known as water glass. When combined with an acid such as hydrochloride (HCl) it forms a salt, which needs to be removed through dialysis or by exchange for H^+ through an acidic ion exchange column (Pierre 2011, p. 22).

$$Na_2O \cdot nSiO_2 + 2HCl + xH_2O \rightarrow nSiO_2 \cdot xH_2O + 2\,NaCl$$

Modern precursors are mostly based on either tetramethyl orthosilicate (TMOS, $Si(OCH_3)_4$) or tetraethyl orthosilicate (TEOS, $Si(OCH_2CH_3)_4$). In both reactions the alkoxides hydrolyze and form silicic acid, which then condenses to SiO_2 (Han 2013, p. 707).

$$\text{TMOS}: Si(OCH_3)_4 + 2H_2O \rightarrow SIO_2 + 4CH_3OH$$
$$\text{TEOS}: Si(OC_2H_5)_4 + 2H_2O \rightarrow SiO_2 + 4C_2H_5OH$$

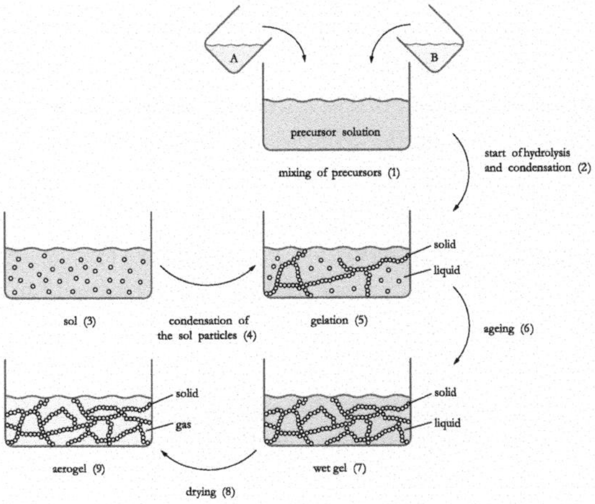

Fig. 4.2 General scheme of aerogel preparation through sol-gel processing (adapted from Hüsing and Schubert 2000, p. 624. Copyright Wiley-VCH Verlag GmbH & Co. KGaA. Reproduced with permission.)

The elimination of the liquid in order to create an aerogel involves further special processes. Aerogels are only produced, when the liquid has been completely replaced by air without altering the solid structure of the wet gel (7). Before drying (8), silica aerogels (9) are often aged in order to strengthen the structural network generated during the sol-gel process. The aging (6) procedure gives extra time for the reactions of alkoxides to finish. Its duration is determined by the thickness of the gel.

4.2.2.2 Supercritical Drying

In supercritical drying, the solvent is shifted into a supercritical state, making the liquid/gas interface in the pores disappear. In practice, this means heating the wet gel inside a closed container so that pressure and temperature exceed the critical temperature and pressure of the solvent (Fig. 4.3). For this the gel is placed in an autoclave in which the temperature gets slowly raised (1). Once temperature and pressure have been kept at values above the liquid's critical point for a certain time, the fluid is gently vented at constant temperature, while the pressure drops. When ambient pressure has been reached the gel gets cooled down to room temperature and the finished aerogel can be removed from the autoclave (2). Since the procedure involves both high operating temperature and pressure (for a TMOS silica aerogel the critical point of ethanol is at 243 °C, 63.8 bar) it bears certain risks (Hüsing and Schubert 2000, pp. 621–646).

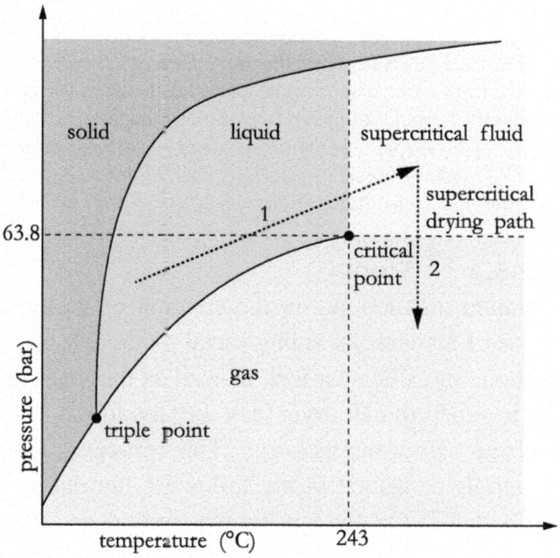

Fig. 4.3 Schematic representation of the principle of supercritical drying based on alcohol (adapted from Finney and Jacobs 2010)

4.2.2.3 Supercritical CO_2 Drying

To increase safety liquid carbon dioxide (CO_2) can be used which allows the operation at moderate pressure and lower temperature (Fig. 4.4). First, the wet gel is pressurized to around 50–60 bar while the temperature is maintained at 5–10 °C (1). Then liquid CO_2 is rinsed throughout the autoclave until all alcohol has been removed from the gel. The temperature is then increased above the critical point of CO_2, which is at 31 °C, 73.9 bar (2). After maintaining the system at this point for a short period the supercritical fluid gets vented off to ambient conditions and the aerogel is formed (3). Since the temperature in this process is much lower than during supercritical drying of alcohols, the process is considerably faster and hence more efficient and cost effective.

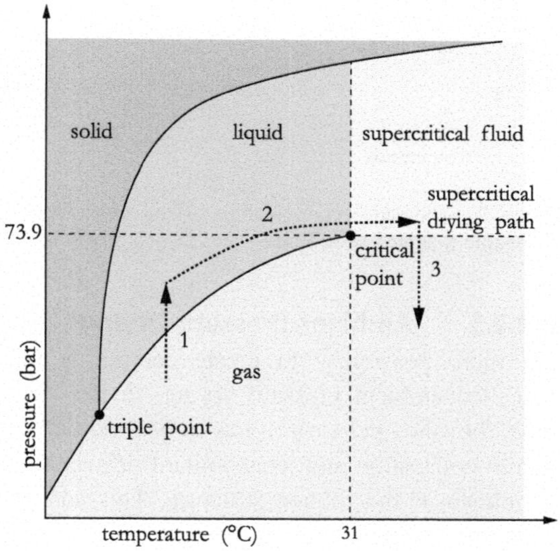

Fig. 4.4 Schematic representation of the principle of supercritical drying based on CO_2 (adapted from Finney and Jacobs 2010)

4.2.2.4 Freeze-Drying

Another technique is freeze-drying (Fig. 4.5). In this method the liquid gel is quickly frozen, and then the pressure is continuously decreased below its sublimation point, where the solvent becomes gas and can be removed from the gel. The resulting material is called cyrogel and its surface area and mesopore volume tends to be smaller than those of aerogels. Known issues of this process for silica sol-gel solutions are that it can be hard to reach the materials freezing temperature and the network might be destroyed due to crystallization of the solvent in the pores. Therefore, freeze-drying is usually only used to create powdered materials.

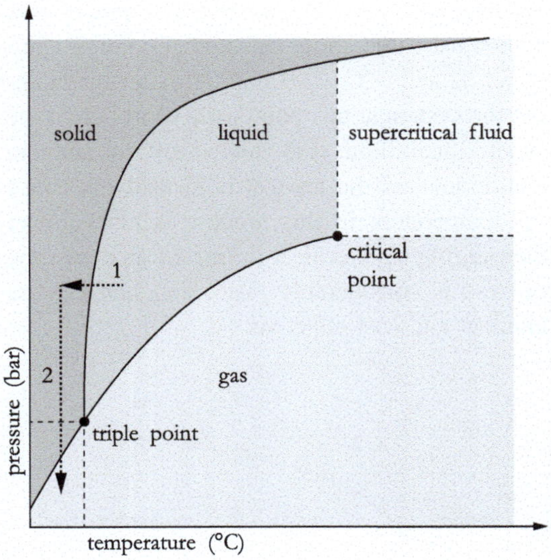

Fig. 4.5 Schematic representation of the freeze-drying technique (adapted from Finney and Jacobs 2010)

4.2.2.5 Ambient Pressure Drying

On more possibility to create aerogel is the so-called ambient pressure drying. In this process, the silica gel surface has to be modified to avoid collapsing and concomitant irreversible shrinkage of the filigree structure. This can be achieved through a number of successive solvent exchanges to increase the strength of the structural network, and hence tolerate a reversible shrinkage (spring back phenomenon) (Pierre and Pajonk 2002, p. 4251). The resulting materials are often called ambigels.

4.2.3 Fabrication

4.2.3.1 Materials and Tools

See Fig. 4.6.

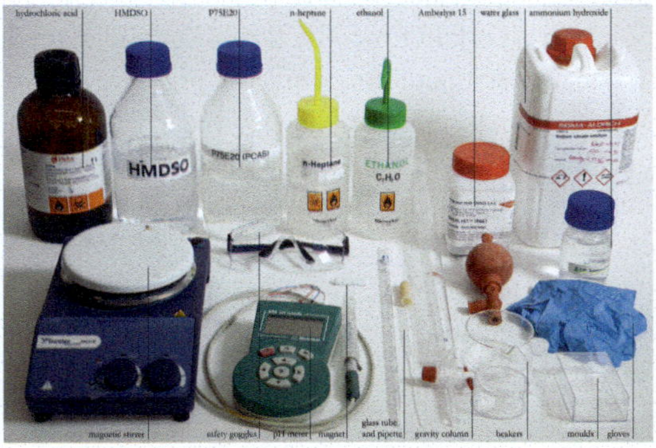

Fig. 4.6 Hydrochloric acid (HCl) 1 mol/L in ethanol, hexamethyldisiloxane (HMDSO)/precursor solution (PEDS-P75E20)/N-heptane/ethanol Amberlyst 15 (wet) hydrogen form ion exchange resin/water glass (sodium silicate) solution/5.5 mol/L ammonium hydroxide/magnetic stirrer/safety goggles/pH meter/magnet(s)/glass tubes/pipettes/gravity columns/several shallow beakers/molds/gloves (Kretzer and Xydis, Aerogel Tutorial 2013)

4.2.3.2 Steps

Detailed instructions on the creation of various types of aerogel, including metal oxide, organic, carbon, and silica aerogel, as well as the making of a supercritical dryer can be found on the website www.aerogel.org. The process and materials described in the following tutorial are for both a TEOS and a water glass precursor. The procedure is based on ambient pressure drying.

TEOS sol (Figs. 4.7, 4.8, 4.9, and 4.10).

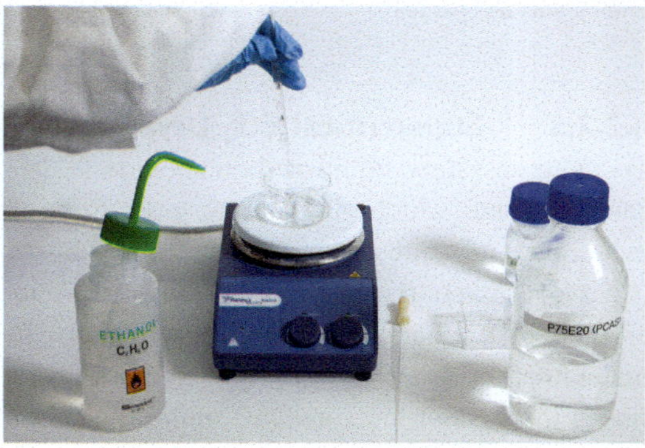

Fig. 4.7 A 20 ml TEOS gel is produced by diluting 7 ml of P75E20 with 12.6 ml of ethanol in a beaker, which is placed on *top* of the magnetic stirrer. The mixture is then stirred for about 5 min at roughly 200 rpm until everything is homogeneously dispersed (Kretzer and Xydis, Aerogel Tutorial 2013)

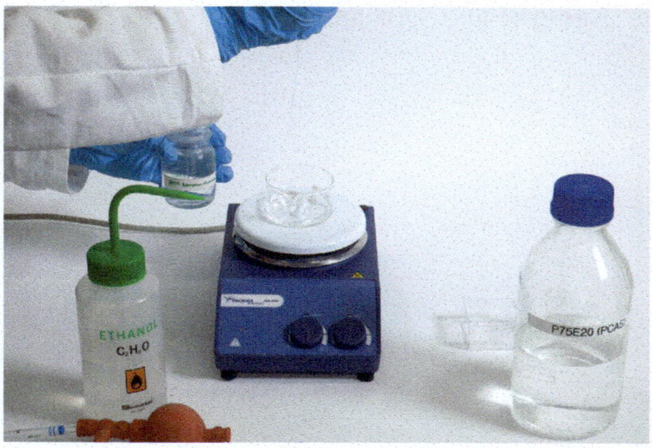

Fig. 4.8 14 drops (0.4 ml) of diluted ammonium hydroxide are added to catalyze the gelation and the mixture is stirred for another 2–3 min (Kretzer and Xydis, Aerogel Tutorial 2013)

Fig. 4.9 Once everything is smoothly dispersed the mixture can be poured into the mold (Kretzer and Xydis, Aerogel Tutorial 2013)

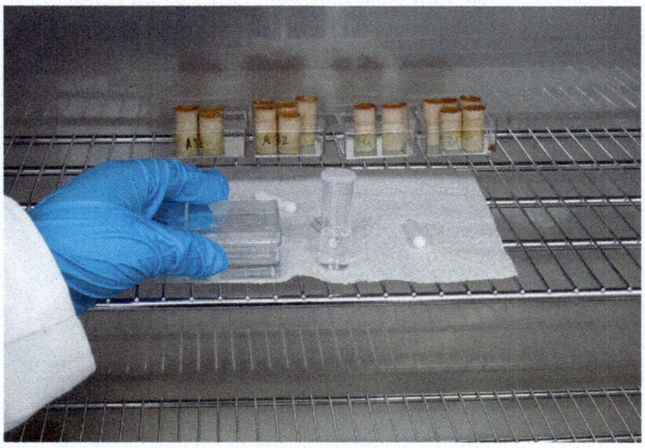

Fig. 4.10 The mold is heated in the oven at 55 °C for 5–10 min until it has formed into a gel (Kretzer and Xydis, Aerogel Tutorial 2013)

Water glass sol (Figs. 4.11, 4.12, 4.13, 4.14, and 4.15).

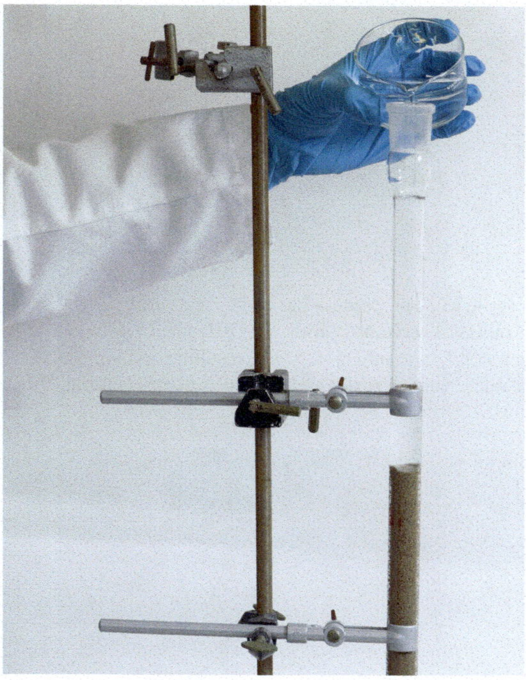

Fig. 4.11 In order to use water glass for aerogel production the pH level of the precursor has to be drastically reduced by performing an ion exchange. This is achieved by pouring 70 ml of Amberlyst 15 hydrogen form into a gravity column that is fixed vertically on a stand. A beaker sitting on *top* the magnetic stirrer is placed beneath the column (Kretzer and Xydis, Aerogel Tutorial 2013)

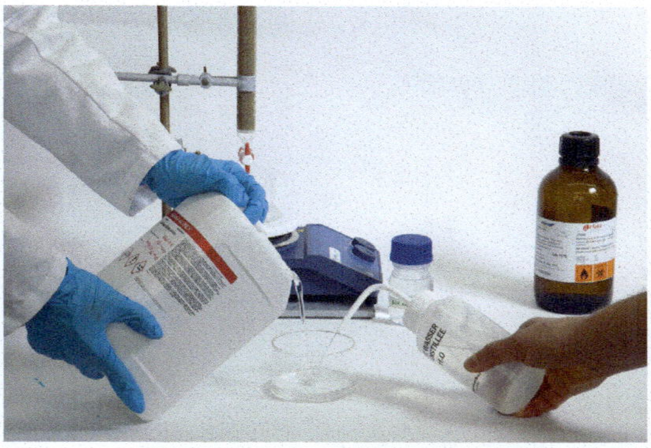

Fig. 4.12 The solution, which is prepared by mixing water glass and distilled water in a ratio of 1:2, is then slowly poured through the column while constantly stirring and measuring the pH level (Kretzer and Xydis, Aerogel Tutorial 2013)

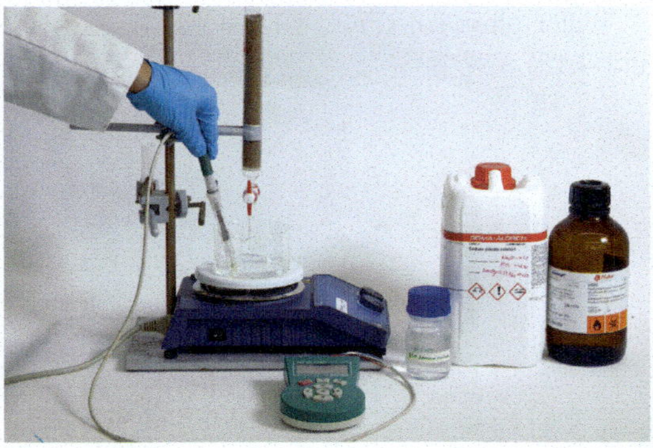

Fig. 4.13 The process has to be repeated by using fresh Amberlyst columns until the pH level is below 2.5, preferably around 1.5–2.5. The resulting sodium silicate is a silicic acid (Kretzer and Xydis, Aerogel Tutorial 2013)

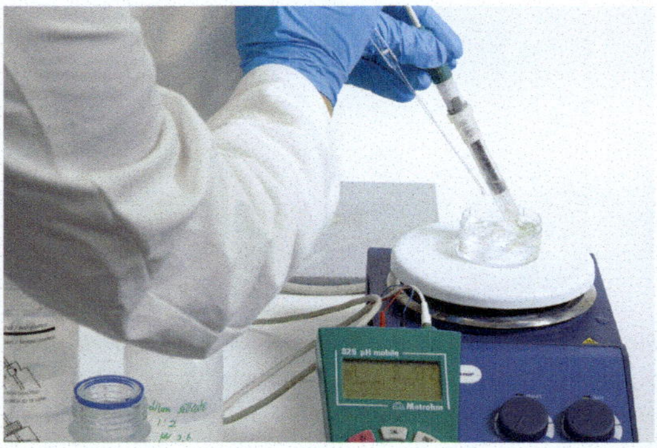

Fig. 4.14 20 ml of the sodium silicate solution are poured into the mold, which stands on the magnetic stirrer. While stirring the mixture for several minutes the pH level needs to be continuously checked. Several drops of ammonium are then added until the pH level has reached 5–5.5 (Kretzer and Xydis, Aerogel Tutorial 2013)

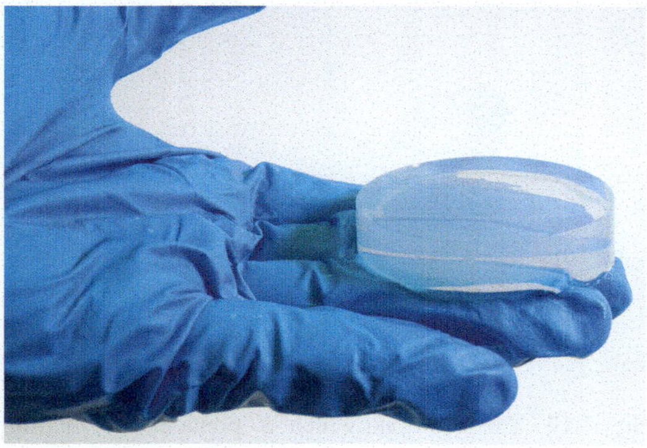

Fig. 4.15 The solution is placed into the oven at 55 °C for 3–5 min until it has turned into a gel (Kretzer and Xydis, Aerogel Tutorial 2013)

Aging (Figs. 4.16 and 4.17).

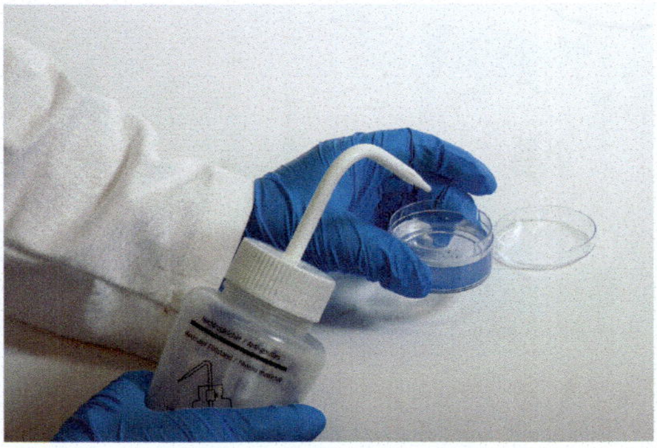

Fig. 4.16 Now the wet gel can be removed from the mold and placed back into a beaker. In order to prevent the occurrence of cracks the TEOS gel has to be covered with excess ethanol, the water glass gel with excess distilled water (Kretzer and Xydis, Aerogel Tutorial 2013)

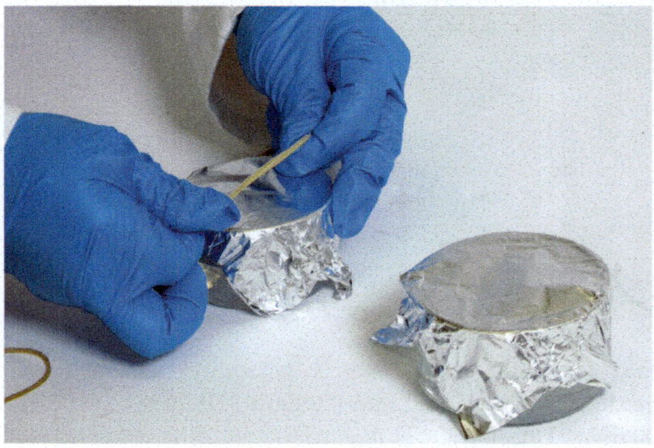

Fig. 4.17 The beaker needs to be closed with a piece of aluminum foil, which gets fixed with a rubber band. The gel is then aged in the oven at 55 °C for 10–12 h (Kretzer and Xydis, Aerogel Tutorial 2013)

Solvent Exchange (Figs. 4.18 and 4.19).

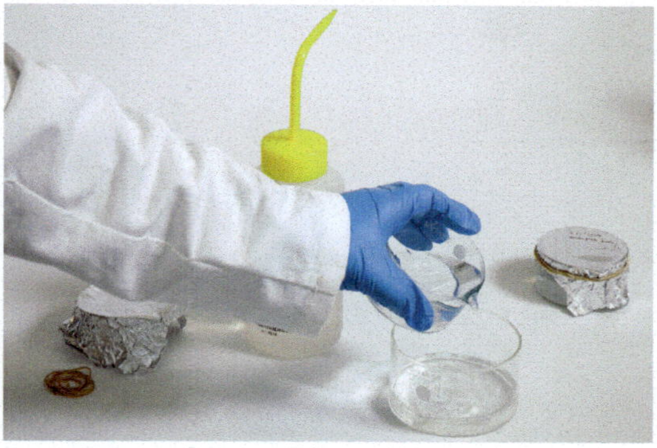

Fig. 4.18 After removing the gel from the oven it has to cool down until it has reached ambient temperature. In case of the TEOS gel, the ethanol solvent is replaced with n-heptane, in case of the water glass gel, the distilled water is swapped for ethanol. The beaker is again covered with the aluminum lid and placed back into the oven for 24 h at 65 °C. For the TEOS gel, the previous step is repeated for a second time and again *left* in the oven for 24 h at 65 °C (Kretzer and Xydis, Aerogel Tutorial 2013)

Fig. 4.19 For the water glass gel the ethanol is now exchanged with n-heptane and the previous step repeated twice (two times 24 h in the oven at 65 °C). From now on the process is the same for both gels (Kretzer and Xydis, Aerogel Tutorial 2013)

Hydrophobization Part I (Figs. 4.20, 4.21 and 4.22).

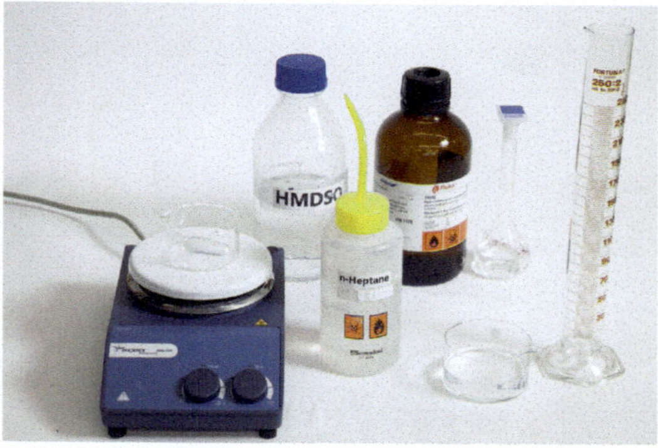

Fig. 4.20 This step marks the most crucial part of the whole process. For 20 ml of gel, the modification solution is prepared by mixing 50 ml of n-heptane, 7.5 ml of HMDSO and 4 ml of a 0.1 mol/L HCl/ethanol solution (Kretzer and Xydis, Aerogel Tutorial 2013)

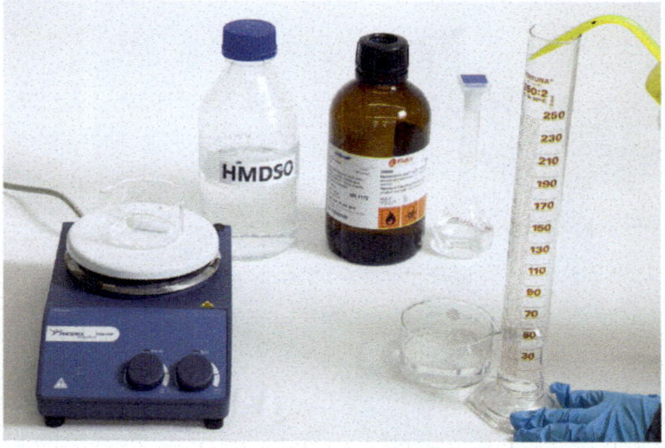

Fig. 4.21 The HCl/ethanol solution is made by filling a 100 ml beaker with 10 ml of 1 mol/L HCl and 90 ml of ethanol (Kretzer and Xydis, Aerogel Tutorial 2013)

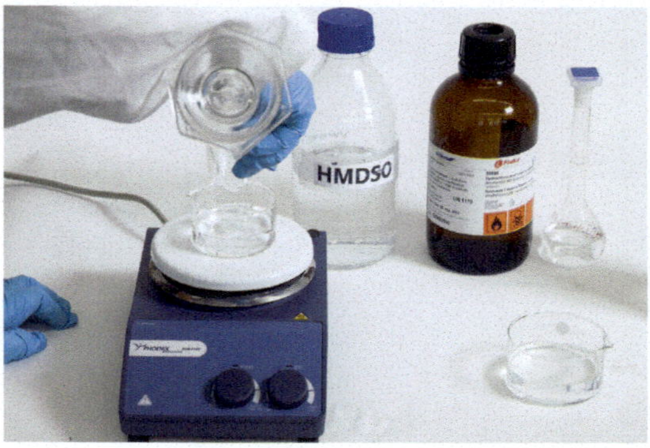

Fig. 4.22 All components are added to a beaker on the magnetic stirrer and stirred for at least 5 min until everything has dispersed equally (Kretzer and Xydis, Aerogel Tutorial 2013)

Hydrophobization Part II (Figs. 4.23 and 4.24).

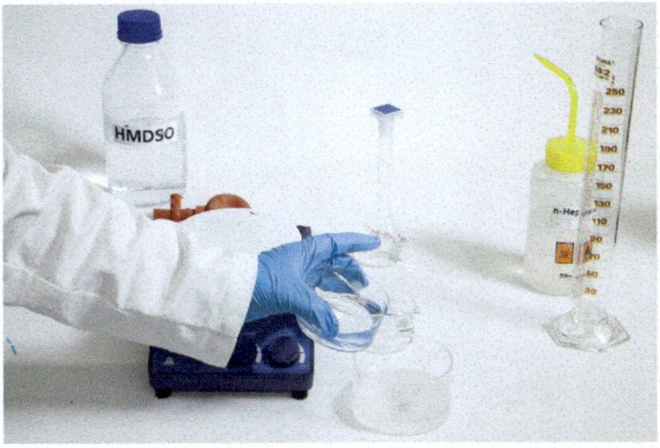

Fig. 4.23 The n-heptane solution of the gel can be discarded and is replaced with the previously prepared modification solution until the gel is completely covered (Kretzer and Xydis, Aerogel Tutorial 2013)

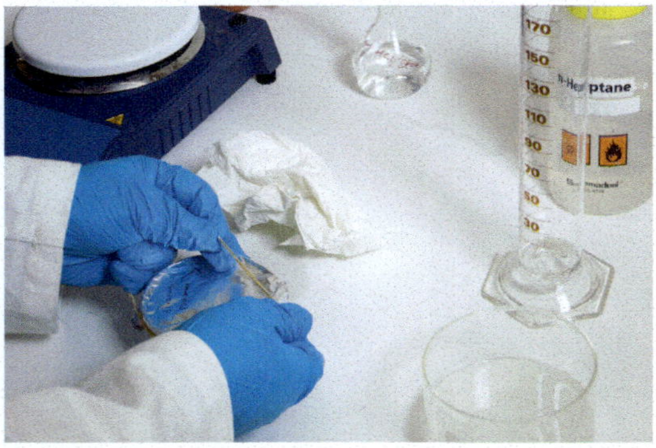

Fig. 4.24 The beaker is again closed with the aluminum lid and placed back into the oven at 65 °C for 24 h (Kretzer and Xydis, Aerogel Tutorial 2013)

Aerogel (Figs. 4.25, 4.26, and 4.27).

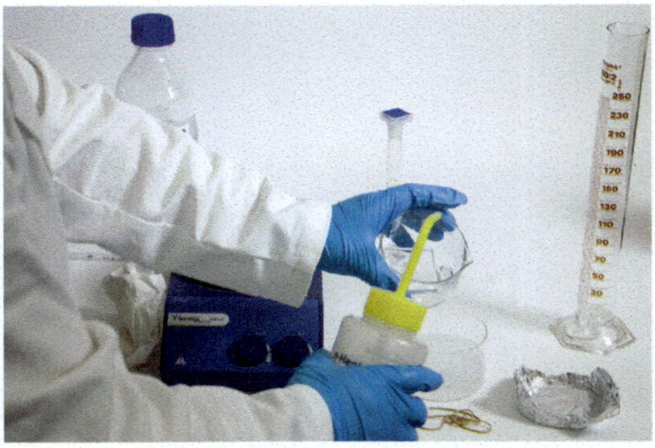

Fig. 4.25 After taking the gel out of the oven it is thoroughly washed and rinsed in n-heptane. The gel without any solvent or liquid is placed back into the beaker and covered with aluminum foil, which is fixed with a rubber band (Kretzer and Xydis, Aerogel Tutorial 2013)

Fig. 4.26 The beaker goes back into the oven at 65 °C in case of the TEOS gel and 55 °C for the water glass gel for two more hours (Kretzer and Xydis, Aerogel Tutorial 2013)

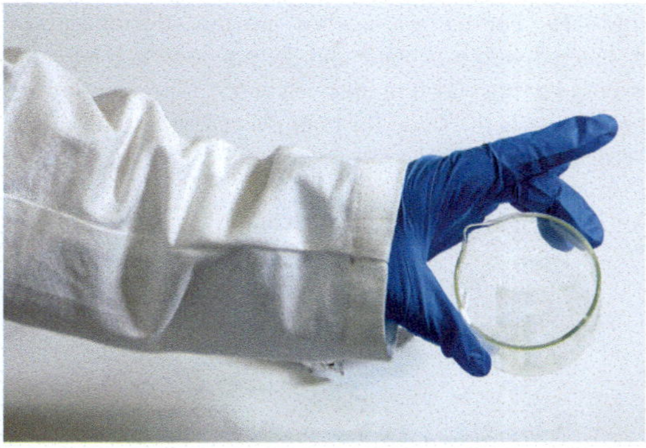

Fig. 4.27 The rubber band is then removed and the gel placed into the oven at 150 °C for two final hours. The result is a (granular) aerogel (Kretzer and Xydis, Aerogel Tutorial 2013)

4.2.4 Applications

Different types of aerogels offer different kinds of applications. Carbon aerogels have a very low electric resistance (<40 mΩ/cm) and are well suited as electrodes in batteries since the surface area per unit is very high and hence more charge can be stored than with conventional capacitors. Metal oxide and organic aerogels are good for low dielectric applications. SiO_2 aerogels can be used as ecologically acceptable insecticides, or as cosmic dust collectors on space crafts.

The most used aerogels and up to date the only kind of true commercial value are silica-based aerogels. They have successfully been applied in space exploration by NASA, as insulation material in freezers, as sound barriers and acoustic insulators or based on their low-refractive index as **cherenkov counters**, which are devices used for electrically charged particle identification in experimental physics (Pierre and Pajonk 2002, p. 4256). The extreme porosity and very large surface area of silica aerogels also makes them suitable for usage as gas filters, (radioactive) waste encapsulation media, and hydrogen fuel storage. They have been proposed as shock absorbing materials and they can be doped with rare earth elements, which allows the production of glass for lasers and radioluminescent light sources (Hüsing and Schubert 2000, p. 21).

For architectural usage silica aerogels have lots of potential as (super)insulating materials, especially for the realization of high-performance windows. Standard double-glazed windows have U-values around 1.0 W/m^2 K, but aerogels sandwiched in between panes of glass can reach U-values below 0.5 W/m^2 K, which is less than triple glazing currently achieves. The production of large, crack-free, and homogeneous plates with high-optical quality, however poses significant challenges and is thus far from being industrially and commercially ready. In comparison to monolithic elements, granulates can be used, which offer various advantages, especially during production, since techniques like subcritical or ambient drying can be applied. Moreover, the effective thermal conductivity of granular beads packed in beds is only slightly larger than that of monolithic structures. Granulates can be infused in blankets, which allow flexibility and geometrical freedom or can be poured like a powder and are thus much easier to handle than the fragile monoliths.

Both granular aerogels and flexible blankets are commercially available and are industrially produced by the North American companies Cabot Corporation and Aspen Aerogels. More recent competitors include Nano Hi-Tech in China and EM-Power from Korea.

Aerogel materials have been used in a number of buildings, including the roof construction of the school Buchwiesen in Zurich, the Dedmond Athletic Centre in Radford, the Talisman Centre in Calgary the British Haley VI Research Station in the Antarctic, the reconstruction of the Florence Mila Borchert Big Cat Country building, a sports hall in Carquefou, France, and the Ruf Maschinenbau factory in Germany (Leydecker 2008, pp. 131–134).

4.3 Bioluminescence

Bioluminescence is the visible creation of light by an organism through a biochemical reaction. It is produced by the sudden transformation of a high-energy molecule into a lower state of energy. The difference in these energy levels is emitted as one photon of light.

Virtually, all living organisms are able to generate light without external photoexcitation but only some of them are bright enough to be noticed. Even human bodies emit small portions of biophotons, 1000 times lower than the sensitivity of the human eye. In 2009 researchers from the Department of Electronics and Intelligent Systems, Tohoku Institute of Technology, Sendai, Japan captured the body glow of five male volunteers throughout three consecutive days using a **CCD** camera (Fig. 4.28). Interestingly, the very weak light emittance has its lowest point at 10 a.m., its peak around 4 p.m. in the afternoon, and is completely independent of body surface temperature and heat emission (Kobayashi et al. 2009, p. 1).

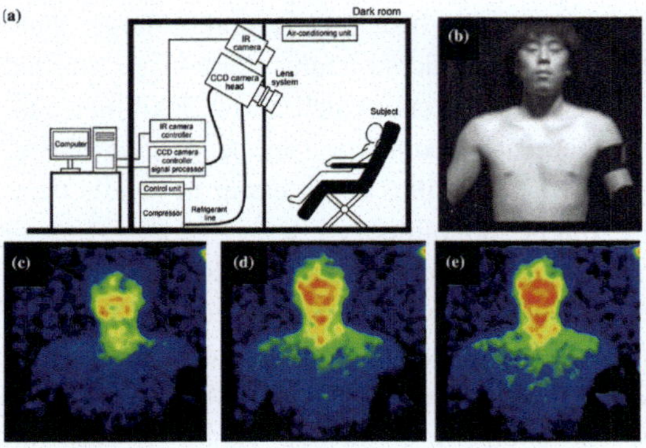

Fig. 4.28 **a** Schematic illustration of experimental setup. **b** Image of the subject under light illumination. **c** Image at 10:10 a.m. **d** Image at 13:10 p.m. **e** Image at 16:10 p.m. (Kobayashi et al. 2009)

4.3.1 History

Some of the oldest written records of bioluminescence observed in nature date back to 1500–1000 B.C. in China, referring to fireflies in folk songs and poetry. The first specific and complete collection of bioluminescent organisms is found in Pliny the Elder's *Naturalis Historia*, which includes fairly comprehensive descriptions of glowworms, fireflies, the luminous mollusk, the purple jellyfish, the lantern fish, luminous mushrooms, and glowing wood (Lee 2008, p. 195). However, it was not until the rise of European Alchemy in the sixteenth and seventeenth centuries until the intellectual foundation allowed real scientific studies in order to understand such phenomena. The oldest book concerned entirely with luminescence, *De Lunariis*, was written by Conrad Gestner, Professor of Natural History and Medicine in Zurich, covering both bioluminescent animals and plants, as well as luminous stones. In 1602, Vincenzo Casciarolo discovered the 'Bolognian Stone,' also referred to as 'Litheophosphorus,' a mineral, today known as Barite (barium sulfate), which was magically able to accumulate light when exposed to the sun and to release it during darkness (Roda 2010, p. 11). In 1647 Bartholin, a Danish physician, published *De Luce Animalium*, describing all luminescent phenomena known at the time, proposing that light was present in every element of life. Robert Boyle, together with his colleague Robert Hooke, carried out various experiments on the effect of air on different forms of bioluminescence. The earliest modern study of bioluminescence was performed on a pyrophorous click beetle by the French pharmacologist Raphael Dubois in 1885. He was subsequently able to extract the two key elements of the bioluminescent reaction and produce light by recombining them artificially (Roda 2010, p. 26). From this, he concluded that the phenomenon was chemical in nature and coined the terms 'luciferine' for the light-producing substance and 'luciferase' for the enzyme that triggered the reaction. The next major step in identifying bioluminescent phenomena came with the advancement of ocean-going vessels in the nineteenth century and especially deep water trawling, which revealed that almost all found specimens were able to produce light. Identifying more than half of all bioluminescent species known until today during the nineteenth century, it was concluded that the bioluminescence phenotype was widely spread in nature (Lee 2008, p. 204).

4.3.2 Material Structure and Operation

Today, it is known that bioluminescence is generally the result of the oxidation of a light-emitting molecule (luciferin) together with a catalyzing enzyme, either a luciferase or a photoprotein, which is a luciferase variant that merges factors required for the light emission in a single unit (Fig. 4.29).

The reaction is highly efficient, turning 98 % of its energy into light (hv), and only loosing 2 % as heat. Both luciferin and luciferase are generic terms rather than the names of certain chemicals, meaning that lots of different substances can lead to the reaction.

$$\text{Luciferin} + O_2 \xrightarrow{\text{luciferase}} \text{Oxyluciferin} + CO_2 + hv$$

It is estimated that bioluminescence has independently evolved at least 40 to 50 times among existing organisms (Haddock et al. 2010, p. 447). The most prominent terrestrial occurrences have been witnessed among insects, mushrooms, and bacteria.

4.3.2.1 Insects

The best-known, earliest documented and widest spread luminescent organisms are luminous beetles, representing roughly 3000 species classified into three families: fireflies, click beetles, and glow-worms. Fireflies and glow-worms are grouped into the luminescent beetle family Lampyridae. They usually light up after the sun has set and display a variety of emission ranging from strong flashes to a steady continuous glow. It is assumed that the light emission among adults forms the basis for sexual attraction and communication among the species. The light is produced in areas of the body referred to as lanterns, which vary in size and location but are usually on the underside of the last abdominal segments. As in most luminescent organisms the chemical reaction is driven by the enzyme luciferase (Luc), which produces light in the combination with **adenosine triphoshpate** (ATP), oxygen (O_2), and magnesium ions (Mg^{2+}) in several steps. During the first step of the reaction firefly luciferin (LH_2) and ATP join with luciferase. In the next stage, an acid anhydride is formed between the luciferin and adenosine monophosphate (AMP), releasing pyrophosphate (PPi). This is then oxidized, generating dioxetanone, which during the final stage through decarboxylation leads to the formation of oxyluciferin in a singlet electronically excited state that quickly converts to its ground state releasing a photon of light with a peak around 530 nm (Day 1968).

$$\mathrm{Luc} + \mathrm{LH_2} + \mathrm{ATP} \xrightarrow{\mathrm{Mg^{2+}}} \mathrm{Luc.LH_2} - \mathrm{AMP} + \mathrm{PPi}$$

$$\mathrm{Luc.LH_2} - \mathrm{AMP} + \mathrm{O_2} \rightarrow \mathrm{Luc.Oxyluciferin^*} + \mathrm{AMP} + \mathrm{CO_2}$$

$$\mathrm{Luc.Oxyluciferin^*} \rightarrow \mathrm{Luc.Oxyluciferin} + hv$$

4.3.2.2 Foxfire

Foxfire is a gentle bioluminescent glow coming from the forest soil or dead wood that can be observed during nighttime. Its source is rapidly growing fungal cells consuming decaying wood. In total over 50 different species of luminescent fungi have been identified, mostly in tropical environments, with the majority belonging to the group of white-spored **Basidiomycetes** (Perry 2007). The biochemical reaction is a two-step process during which chemical energy resulting from respiration or photosynthesis is loaded onto a fungal luciferin molecule and is then, in combination with oxygen and luciferase, transformed into water, low energy luciferase, and a photon of light (Coder 1999, p. 2). Both the involved luciferans and luciferases remain largely unidentified.

$$\mathrm{L} + 2\mathrm{NADH} \xrightarrow{\mathrm{oxydase/reductase}} \mathrm{LH_2} + 2\mathrm{NAD} +$$

$$\mathrm{LH_2} + \mathrm{O_2} \xrightarrow{\mathrm{luciferase}} \mathrm{L} + \mathrm{H_2O} + hv$$

The produced light is usually in the greenish spectrum and very dim with a maximum around 520–530 nm. While, the true purpose of fungal bioluminescence is unknown it is generally believed that the light attracts insects to spread the mushroom's pores or acts as a deterrent toward nocturnal animals.

4.3.2.3 Luminous Bacteria

Bacterial bioluminescence is the widest distributed source of organic light emittance. Luminescent Bacteria are all **gram-negative**, non-spore forming, and motile. While most of them can live independently, the majority is found in symbiotic or parasitic relationships with host organisms. By now 11 species have been classified which can be divided into four genera: Photobacterium, Vibrio, Alteromonas, and Photorhabdus (Charrier et al. 2010, p. 179). Bacteria belonging to the Photobacterium genus are in most cases light organ symbionts of marine animals while the Vibrio species exist both as free-living and symbiotic types. In bacterial bioluminescence the luciferase enzyme is composed

of reduced flavin mononucleotide, molecular oxygen, and long-chain fatty aldehyde. The reaction leads to the oxidation of $FMNH_2$ and aldehydes and results in the emission of blue/green light with a maximum around 490 nm.

$$RCHO + FMNH_2 + O_2 \xrightarrow{\text{luciferase}} RCOOH + FMN + H_2O + h\nu$$

4.3.2.4 Deep-Sea Bioluminescence

While bioluminescence can be observed among a large variety of organisms, most actually exist in the ocean and in particular the twilight or disphotic zone, which generally extends from about 200–1000 m depth (Wilson 2007).

Even though the divergence from marine to terrestrial bioluminescence is not fully understood and mostly assigned to the ocean's comparatively stable environmental conditions that is favorable for a long uninterrupted evolutionary history, it is estimated that about 90 % of animals living below 500 m depth have luminescent capabilities (Fig. 4.29) (Frank and Widder 2002).

With the seawater not only absorbing red, orange, and yellow sunlight but also significantly changing its intensity about tenfold every 75 m, most of the visible light that reaches this area is in the blue/green spectrum (Fig. 4.30). Subsequently, the majority of marine organisms are only sensitive to blue light, and hence luminescence in a spectrum of 440–479 nm.

Fig. 4.29 Moon jellyfish (Aurelia aurita), as exhibited in many public aquariums, are often used to visualize deep-sea bioluminescence. However they are actually not able to produce their own light, but due to their translucent body only reflect and disperse light shining upon them (Kretzer, Moon Jellyfish 2014)

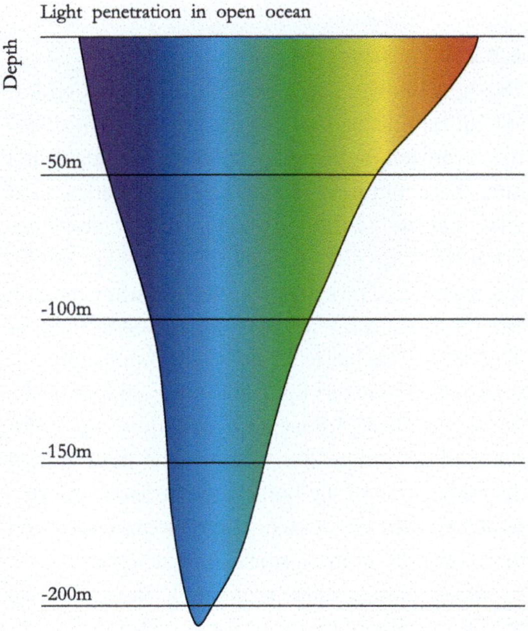

Fig. 4.30 Light penetration in open ocean and spectrum of visible light (adapted from Kyle Carothers, NOAA-OE: Light penetration in open ocean, http://oceanexplorer.noaa.gov/explorations/04deepscope/background/deeplight/media/diagram3.html)

Oceanic bioluminescence is mainly created by four distinct luciferins, bacterial, dinoflagellate, ostracod, and coelenterazine and can be observed among a variety of species with dinoflagellates representing the most commonly encountered organism next to fireflies. In large accumulations they can form spectacular natural phenomena but also potentially toxic red tides (Haddock et al. 2010, p. 455). The reason for their bioluminescent behavior is still not entirely clear but might be used as a kind of burglar alarm, making the feeding fish visible and vulnerable to attack from larger predators. Other reasons for bioluminescence in marine environments include:

- Bright flashes to be used as defense by startling and intimidating approaching predators.
- Ventral photophores match the dim light coming from the surface. This counter illumination makes a dark body disappear from below.
- Glowing lights can work as a lure and attract potential prey.
- Bright colors and patterns might be used as aposematism and repel predators.
- Some fish produce red light, which allows them to illuminate their prey without being visible.
- Signaling patterns are used as means of reproduction-related interspecific communication.
- In some cases light emission might not have a specific function and simply constitute a metabolic by-product.

4.3.2.5 Synthetic Bioluminescence

In 1986, scientists at the University of California successfully transferred the Firefly gene, which is responsible for the insect's ability to chemically glow, into a tobacco plant (Ow et al. 1986, p. 856). In 2003, Professor H.J. Tsai at National Taiwan University developed a method to make otherwise colorless fish shine neon green using a protein extracted from jellyfish (Dean 2003). These and other recent advances gave rise to the 'Glowing Plant Project,' an open source initiative, which attempts to openly spread the technology in the public domain (Glowing Plant). The idea was initiated in April 2013 by a small group of California-based **biohackers** who organized a **Kickstarter** campaign to fund their plan to synthetically engineer and produce a glowing plant and exceeded their initial goal of 65,000 USD by more than seven times. The plant, which is in the long-run ought to become a sustainable alternative to light bulbs and street lamps instigated intense controversies concerning ethical, moral, and environmental issues due to its unregulated kind and the unknown implications a genetically engineered plant could pose to nature. In October 2013, the UK-based experimental entrepreneur Charlie Francis presented a glow in the dark jellyfish ice cream using calcium activated proteins that react when they are agitated (Kinder 2013).

4.3.3 Fabrication

Even though bioluminescence is a natural phenomenon it can be recreated in an artificial environment. The easiest way to do so is by breeding either a bacterium, for example, Vibrio fischeri, or by growing dinoflagellates, like Pyrocystis lunula or Pyrocystis fusiformis. Generally, bacteria emit a more continuous glow, while dinoflagellates produce a bright flash of light, when stimulated. The following tutorial describes the controlled growth of Pyrocystis lunula.

4.3.3.1 Materials and Tools

See Fig. 4.31.

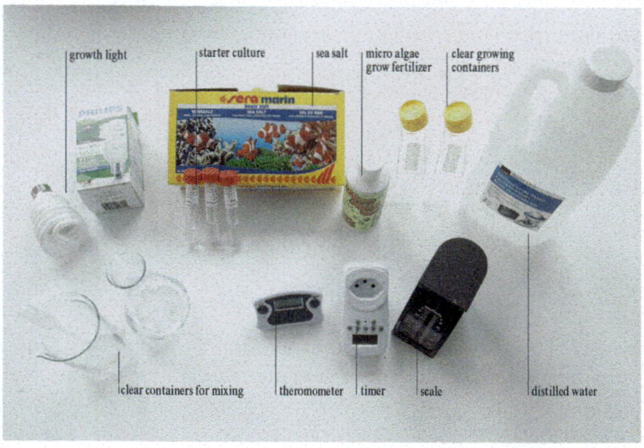

Fig. 4.31 Grow light(s) (T8 or T5 with a 9.56 light spectrum are recommended)/sea salt/distilled water/starter culture(s) of Pyrocystis lunula/micro algae grow fertilizer/clear and clean growing containers (large, flat containers work best)/dropper or syringe/clear containers for mixing the solution/thermometer/timer/scale, preferably digital (Kretzer, Bioluminescent Tutorial 2013a, b)

4.3.3.2 Steps

Growth medium (Figs. 4.32 and 4.33).

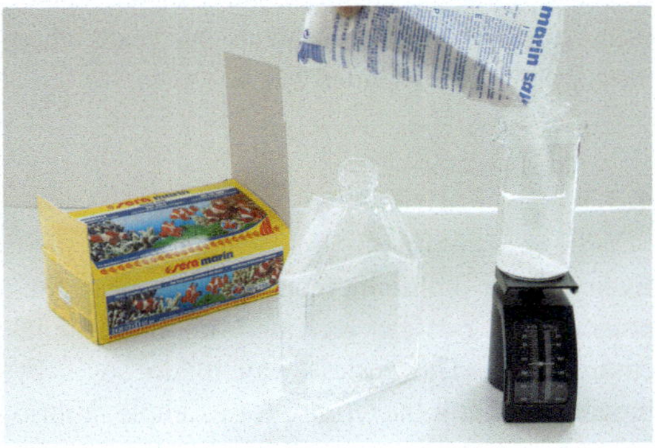

Fig. 4.32 The salt-water solution, which will become the grow medium, is made by mixing 33 g of sea salt, which can be obtained in any pet-shop, with 1 L of distilled water. It is easier to mix a larger batch and then split it up in smaller quantities. The amount of salt has to be measured precisely, since too much salt might harm the organisms (Kretzer, Bioluminescent Tutorial 2013a, b)

Fig. 4.33 Cell culture bottles, which are sterile and clean, work well as growing containers. After mixing the medium it should be shaken thoroughly until the salt has dissolved completely (Kretzer, Bioluminescent Tutorial 2013a, b)

Growth Solution (Figs. 4.34, 4.35 and 4.36).

Fig. 4.34 The growth medium will be equally distributed among the growing containers (Kretzer, Bioluminescent Tutorial 2013a, b)

Fig. 4.35 The starter cultures come in 10 ml and should be added to 50 ml of growth medium and 0.025 ml of micro algae grow (Kretzer, Bioluminescent Tutorial 2013a, b)

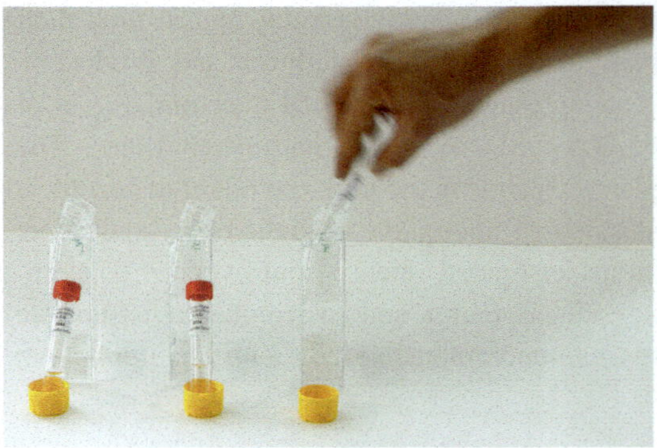

Fig. 4.36 Both starter culture and solution should have the same temperature before mixing (between 18 and 23 °C) (Kretzer, Bioluminescent Tutorial 2013a, b)

Growth (Figs. 4.37 and 4.38).

Fig. 4.37 The containers should be placed in a dark place together with the grow lights. It is necessary to precisely control and time how much light they receive. The timer needs to be set so the organisms remain in a constant cycle of light (14 h) and darkness (10 h). The cultures can be programmed to light up during daytime. During their night cycle they will not emit any light. While growing the cultures should be observed carefully and gently shaken every day to prevent sediment collection on the *bottom* of the container (Kretzer, Bioluminescent Tutorial 2013a, b)

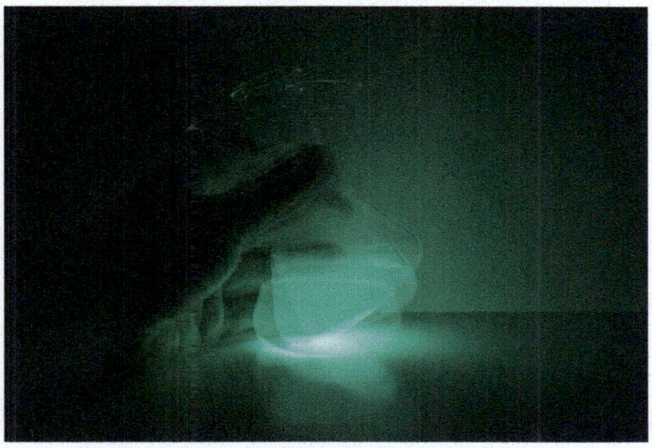

Fig. 4.38 Once the cultures start emitting light they might be split into two batches. This should be done at least every three to four weeks in order to provide sufficient nutrition. The dinoflagellates will only flash when disturbed or stressed and exclusively during their night cycle. Too much stress will tire and harm them (Kretzer, Bioluminescent Tutorial 2013a, b)

4.3.4 Applications

Bioluminescent research targets many areas, including genetic engineering, biomedical applications, or military use. Inspired by the symbiotic relationships of luminous bacteria and their host organisms, studies in architecture and design investigate the potential application of similar bacteria in a spatial or material context.

Students from the University of Cambridge participated in the 2010 International Genetically Engineered Machine (iGEM) competition and proposed to reduce the amount of power used for lighting, which accounts for 8 % of our use of electricity, through bioluminescent trees that photosynthesize during the day and emit light at night. They proved their concept by building a bacterial lamp as well as printing an emergency exit sign containing strains of bioluminescent **E. coli** (Eglowli Cambridge).

Eduardo Mayoral, PhD candidate at GSAPP Columbia University, won the third price of the 2011 Holcim Award for Sustainable Construction with his vision 'Bioluminescent devices for zero-electricity lighting.' The project's aim was to develop self-glowing devices through the manipulation of bioluminescent microorganisms, which could be used for ambient public lighting, natural park illumination, billboards, screens, highway posts and signs, and many other uses (Holcim).

As part of their Microbial Home Probe project, which consisted of a domestic ecosystem that challenged conventional design solutions regarding energy, cleaning, food preservation, lighting and human waste, Philips Design presented the 'Bio-Light' in 2011. The wall of glass cells used living bioluminescent bacterial cultures that were fed with methane and composted materials and emitted a soft green light (Philips).

The German designer Nicola Burggraf has extensively worked with bioluminescent algae and created a number of sound and touch responsive ambient environments that have been publicly exhibited, such as 'Bioluminescent Field' at the Luminale, Palmengarten Frankfurt in 2010 or 'Interference' at the Luminale in 2012 (Burggraf).

The 2015 winner of MoMA's PS1 Young Architect's program Andres Jaque planned to build a towering labyrinth of water pipes, called 'Cosmo' in order to create an awareness among visitors about global water consumption. The machine was intended to purify 11,500 L of water over the course of a few days, filtering it through a multitude of natural and synthetic ecosystems including algae, plants, ultraviolet rays, and waterfalls to cleanse it from pollution and enrich it with oxygen. Moreover, Jaque wanted to add bioluminescent microorganisms, which would light up the courtyard during nighttime (Dezeen 2015).

4.4 Bioplastics

A Bioplastic or biopolymer is a substance that is composed of renewable organic biomass sources, such as starch, cellulose, or sugar (Goodall 2011, p. 1). Because of its biological origin, it is inherently biodegradable, which means that it can easily be broken down into CO_2, water, energy, and cell mass with the aid of microbes, rendering it largely carbon neutral (Lorcks 2006, p. 5). On top of its ecological advantage to standard plastic, which is largely derived from petrochemicals and can take hundreds of years to degrade, it helps to conserve fossil raw materials and the dependency on mineral oil.

When looking at plastic there are generally two main types to consider:

- A thermoset is a plastic that, after curing, remains insoluble, infusible, and essentially solid. It is usually liquid or malleable prior to its curing process but cannot be reheated or melted back, since it would reach its decomposition temperature before the melting point. This means that a thermoset cannot be recycled, except as filler material but is generally much stronger and durable than a thermoplastic material. Examples include polyurethanes (PU), Bakelite, duroplast, and polyester fiberglass systems, melamine, and epoxy resins or polyimides (PI).
- A thermoplastic can be softened and hardened repeatedly through heating and cooling. In its softened state it can be formed into any desired shape by molding or extruding techniques. Thermoplastics make about 90 % of all plastics produced today and include acrylic (PMMA), nylon, polyethylene (PE), polypropylene (PP), polystyrene (PS), polyvinyl chloride (PVC), and Teflon (PTFE).

In order to encourage plastic recycling and simplify separation, the Society of the Plastic Industry (SPI) introduced the Resin Identification Code (RIC) system in 1988, a set of symbols assigned to and printed on the most common types of plastic containers. Correct recycling is important since a single item of the wrong type could ruin the whole batch, nevertheless the involved energy consumption is enormous and generally shows a negative ecobalance (Avérous and Pollet 2012, p. 14). More problematic than proper recycling however is the plastic that is not being reused, but rather litters and disfigures coastlines and oceans and poses a danger on marine life. Since a lot of this plastic also leaks small amounts of pollutants, such as **endocrine disruptors**, into the environment, it can have serious impacts on animals and humans.

For these reasons a continuous interest in replacing traditional plastic with biodegradable materials prevails, particularly for products with a short life span, such as packaging, catering, surgery, hygiene, or engineering applications.

4.4.1 History

The beginnings of bioplastic cannot be clearly dated, but natural resins, such as amber, shellac, or **gutta percha**, were already used during Roman times and in the Middle Ages. A different technique was developed by Native Americans, who were boiling animal bones, horns, and hooves in water and then molded or pressed them into sheets to make ladles and spoons.

Commercialization of bioplastic started in the middle of the nineteenth century. In 1845, Swiss-German chemist Christian Friedrich Schönbein prepared a strong, transparent and waterproof cellulose derivative from paper (cellulose nitrate), which could be shaped into a variety of forms. In 1846 Louis Ménard discovered that drying a solution of cellulose nitrate soaked with ethanol, called collodion, would lead to a tough, elastic, and waterproof solid material. In 1848, J. Parker Maynard spread collodion on open wounds and found out that it would dry into an air—and watertight film that fostered the healing process. At the 1862 building exhibition in London the English inventor Alexander Parkes presented a pressure-molded product from collodion that he called Parkesine and which he brought to the market in 1866. However, only 2 years later the company failed mostly due to the material's flammability, premature production, and the use of substandard raw materials. In 1869, John Wesley Hyatt Jr. patented the use of collodion for coating billiard balls in order to develop an alternative to expensive and precious ivory. Together with his younger brother Isaiah Smith Hyatt they patented a process for plasticizing cellulose nitrate with camphor in 1870 and consequently opened their first factory to mass-produce what they called celluloid. Celluloid became popular in 1882 to create photographic films and later for making the first motion-picture films. At the end of the nineteenth century Adolf Spitteler, a Bavarian chemist, invented moldable casein plastic, which was durable, corrosion-resistant and pleasant in its appearance. The production of casein plastic became a strong industry at the beginning of the twentieth century. Research and advancements into organic chemistry however soon led to a shift toward the use of fossil fuels for plastic production and in turn the discovery of Bakelite in 1907 and many other plastics during the 1920s. In the 1910s, Henry Ford began experimenting with converting soybeans into plastic for the use of automobile parts. In 1941, he exhibited a prototype car, whose body consisted of 14 pressure-molded panels made from soybean plastic. Further developments were however stopped with the beginning of World War II. In 1923, the industrial production of cellophane began, the only bioplastic that until today survived the growth of the synthetic plastic industry. Cellophane is a sheet material derived from the abundant cellulose, which is found in plant cell walls and makes up around 40 % of all organic matter. The commercial use of cellophane as transparent, waterproof packing material started in the early 1930s, its production had its peak in the 1960s but today it has been largely replaced with synthetic materials, like polypropylene (Stevens 2002, pp. 105, 107–118).

Since the 1980s research and development into bioplastic has resumed and is constantly increasing, mostly due to a growing environmental awareness and a fear of dependence on nonrenewable resources. However, until today bioplastic has not found much use to replace synthetic plastic materials, largely due to its high manufacturing costs and often inferior performance. Nevertheless, the market is in rapid expansion with an annual growth rate of 10–20 % (Avérous and Pollet 2012, p. 13), a proposed 500 % increase in production capacity until 2016, and an estimated volume on the plastic market of 20 % by 2020 (Washam 2010, p. 10).

4.4.2 Material Structure and Synthesis

Depending on the development of the bioplastic synthesis process, different classifications of the various biodegradable polymers exist. Figure 4.39 displays a system based on four different categories. Polymers from biomass (1), such as starch or cellulose, polymers based on microbial production (2), like the polyhydroxyalkanoates, polymers synthesized from monomers obtained from biological resources (3), e.g., the poly(lactic

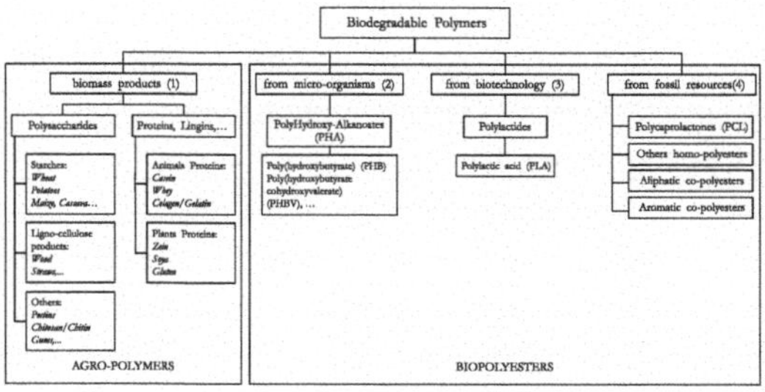

Fig. 4.39 Classification of main biodegradable polymers (adapted with kind permission from Springer Science+Business Media from Avérous and Pollet 2012, p. 16, Fig. 2.1)

acid), and polymers derived from fossil resources (4) (Avérous and Pollet 2012, p. 16).

Biodegradable polymers can be split further into two main classes: the agro-polymers and the biodegradable polyesters. Agro-polymers, are generally extracted from plants, are compostable and renewable and often exhibit a hydrophilic character. The main types include polysaccharides and proteins. Biopolyesters or biodegradable polyesters include the most common types of bioplastic, the polylactides, like PLA, and the polyhydroxyalkanoates, like PHA and PHB.

4.4.2.1 Polysaccharides

About 80 % of today's bioplastic production is based on starch, a natural compound, produced in plants during their photosynthesis. Starch is derived from maize, wheat, potatoes or tapioca and can directly be processed into bioplastic. Since thermoplastic starch (TPS) is soluble in water, articles made from starch swell and deform when exposed to moisture and are therefore mainly used in the pharmaceutical industry. This problem can however be solved by chemically modifying the thermoplastic starch into a different polymer, which broadens the spectrum of possible applications (Peters 2011, p. 37).

In addition to starch other abundant polysaccharides include chitin, cellulose, agar, carrageenan, pectin, alginate and lignin. Chitin-based material is commercially produced as film or fibers. Bioplastic made from cellulose can reach light permeability of up to 90 %. The best-known examples are cellulose acetate (CA) and cellulose triacetate (CTA). Lignin represents the second most common biopolymer found in nature, next to cellulose, making up about 20 % of all organic matter. It serves as the stiffening substance in the outer layer of every plant cell and accounts for about one third of a tree's material. It is extracted through a boiling process from wood shavings and fibers and enhanced with methanol and hydrochloric acid to form a resin-like substance. Polymer blends from lignin exhibit good mechanical properties and a high degree of rigidity.

4.4.2.2 Proteins

Protein is produced by plants, animals, and bacteria. Various proteins have been evaluated as biodegradable polymers, however only a few have been industrially produced, mostly due to high manufacturing cost and low performance. Soy protein, zein, gelatin, and casein are among the most promising materials. Plastic from soy protein can be extruded and injection molded. Zein has thermoplastic properties, is water insoluble, and can withstand microbial attacks. It can be processed into fibers that are strong, washable, dye-able, and grease resistant and is thus used in cloth manufacturing. Gelatin-based plastic is mostly used for drug and vitamin encapsulation or other biomedical applications, like artificial skin, due to its compatibility in physiological environments (Stevens 2002, pp. 90, 124).

4.4.2.3 Polyhydroxyalkanoates

Polyhydroxyalkanoate (PHA) is mainly produced by bacterial fermentation of sugar or lipids. The bacteria produce a range of different PHA molecules, which can consist of more than 150 types of monomers, leading to materials with extremely varying properties (Washam 2010, p. 11). Through the combination with other polymers, enzymes or inorganic compounds, the materials can be further diversified. PHA is more ductile but less elastic than other bioplastic and is largely used in the medical industry. The primary biopolymer of the PHA family is the polyhydroxybutyrate homopolymer (PHB) also known as polyhydroxbutric acid. PHB is a **biocrude**, nontransparent, crystalline polyester with a high melting point, very similar to polypropylene. Since it does not dissolve in water and is resistant to hydrolytic degradation it proposes a good alternative to most other biodegradable plastics, which are usually moisture sensitive. PHB production, a transformation of glucose, cornstarch, or wastewater through certain bacteria, is currently developed to an industrial scale. Its mechanical properties can be tailored to exhibit outstanding flexibility and toughness and it is biodegradable without residue.

4.4.2.4 Polylactides

Polylactic Acid (PLA) is a biocrude plastic which can be produced either by bacterial fermentation of starch from crops into lactic acid that is then polymerized or by chemically fermenting viscous sugar syrup. It is a transparent, colorless, and shiny plastic whose properties are similar to common petroleum based plastics like polyethylene (PE), polypropylene (PP), or polystyrene (PS). It is water-repellent, printable, and breathable and is used in a number of applications including computers and mobile phone casings, biodegradable medical implants, foil, tins, cups, bottles, and other (food) packaging. Due to its good market presence and comparatively low price it represents one of the most promising bioplastics and is expected to overtake the use of PET (Howes and Laughlin 2012, p. 159). Depending on the particular requirements it can be designed to either biodegrade rapidly or to remain stable for several years. Since some PLA soften at relatively low temperatures its mechanical resistance can be strengthened with fibers.

4.4.2.5 Polycaprolactones

Polycaprolactone (PCL) is a biodegradable polyester that is based on petroleum resources and chemically produced from synthetic monomers. PCL is soft at room temperature, has a low melting point, and is commonly used as a PVC solid plasticizer or in PU applications. Due to its biodegradability it has also been applied in biomedical or environmental applications (Avérous and Pollet 2012, p. 32). Based on its soft character and easy handling it is a prominent material in the hobbyist market.

4.4.2.6 Polyamides

Polyamide 11 (PA11) or Nylon 11 is a biopolymer that is produced from vegetable oils but not biodegradable. Its properties are similar to other polyamides but it has a much lower environmental impact, needs less nonrenewable resources, and provides better thermal resistance. It is generally weaker but more resilient than other types of Nylon and gets often reinforced with fibers to increase heat stability and durability (Carson et al., p. 3). Its use lies mainly in high-performance applications like automotive fuel lines, pneumatic airbrake tubing, electrical anti-termite cable insulation, oil and gas flexible pipes and control fluid umbilicals, sports shoes, electronic device components, and catheters.

The following Table 4.1 compares a number of bioplastic materials, like cellulose acetate (CA), polyhydroxyalkanoate (PHA),

Table 4.1 Comparison of various (bio)polymers according to selected physical, mechanical, thermal, electrical, and other properties (Hough and Dolbey 1995, Goodfellow)

Properties	CA	PHB	PLA	PCL	PA11	PMMA	PP	PET
Density (g/cm^3)	1.28–1.3	1.18–1.25	1.21–1.25	1.1–1.146	1.04–1.06	1.17–1.19	0.9–0.91	1.3–1.4
Max. operating temperature (°C)	60	95	50		70	50	100	60
Melting point (°C)	170–240	168–182	150–162	58–65	200–260	220–240	130–166	260–280
Thermal conductivity at 23 °C (W/m * K)			0.12–0.13	0.17–0.18	0.32–0.35	0.17–0.19	0.1–0.22	0.15–0.4
Tensile strength (MPa)	30	20–40	50	10	25–48	70	23–26	55–80
Elongation at break (%)	35	6–10	6	300	50	2.5–4	80	300
Water absorption (%)	2.2	0.75	0.68		0.3	0.3	0.04	0.16
Dielectric strength (MV/m)	11		12	16	15	20	18	40
Flammability (UL 94)	HB		V0	V0	V2	HB	HB	HB

polyhydroxybutyrate (PHB), polylactic acid (PLA), polycaprolactone (PCL), and polyamide (PA 11) with prominent traditional plastics, such as polymethylmethacrylate (PMMA), polypropylene (PP), and polyethylene terephthalate (PET). The materials are compared according to their density, maximum operating temperature, melting point, thermal conductivity at 23 °C, tensile strength, elongation at breaking, average water absorption, dielectric strength, and flammability (based on UL 94, the standard for safety of flammability of plastic materials for parts in devices and appliances testing).

4.4.3 Fabrication

Even though there exist a large variety of different types of bioplastic the fabrication procedure is usually very simple. The following process will produce a thin sheet (roughly A3 in size) of a gelatin-based bioplastic, which can be post-processed using a laser cutter. In contrast to self-made bioplastic produced from cornstarch, gelatin has proven to be a more durable and lasting base substrate. The desired strength and flexibility can be adjusted depending on the mixing ratio.

4.4.3.1 Materials and Tools

See Fig. 4.40.

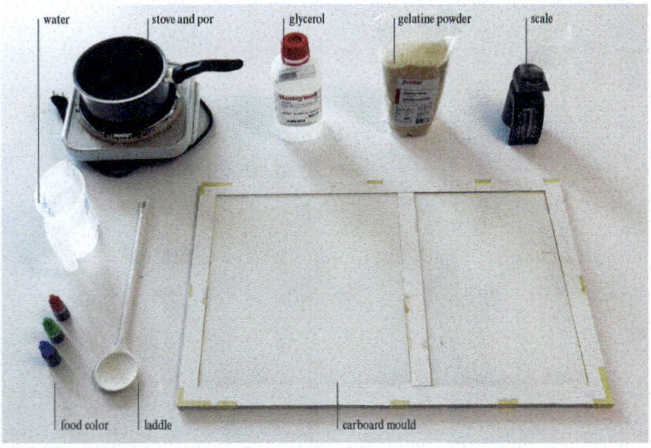

Fig. 4.40 Glycerol/gelatin powder/water/mold for desired shape/food color/aluminum foil/stove/cooking pot/scale/measuring jug (100 ml + 500 ml)/spoons, cups, eggbeater/thermometer (Roman and Pantazis, Bioplastic Tutorial 2012)

4.4.3.2 Steps

Mold making and Preparation (Figs. 4.41 and 4.42).

Fig. 4.41 The mold can be made from cardboard or any other material. It is however essential that the form has no holes or weak parts to prevent leaking of the liquid bioplastic. More complex shapes can be designed, but depending on the thickness the material can take very long to dry fully (Roman and Pantazis, Bioplastic Tutorial 2012)

Fig. 4.42 The work area needs to be clean and all tools and materials should be at hand (Roman and Pantazis, Bioplastic Tutorial 2012)

Mixture (Figs. 4.43, 4.44 and 4.45).

Fig. 4.43 240 ml of cold water is poured into the pot (Roman and Pantazis, Bioplastic Tutorial 2012)

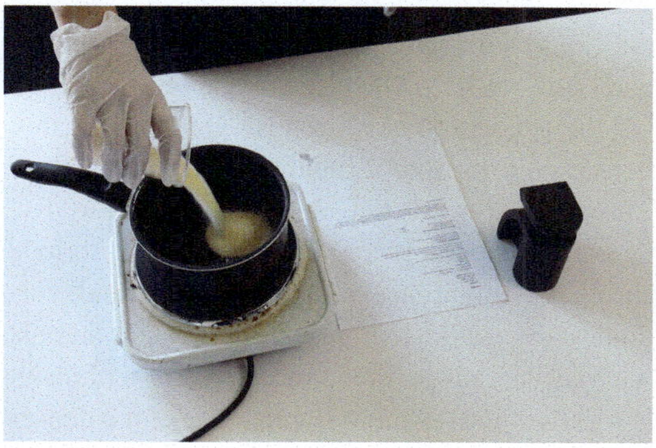

Fig. 4.44 The water is then mixed with 48 g of gelatin powder (Roman and Pantazis, Bioplastic Tutorial 2012)

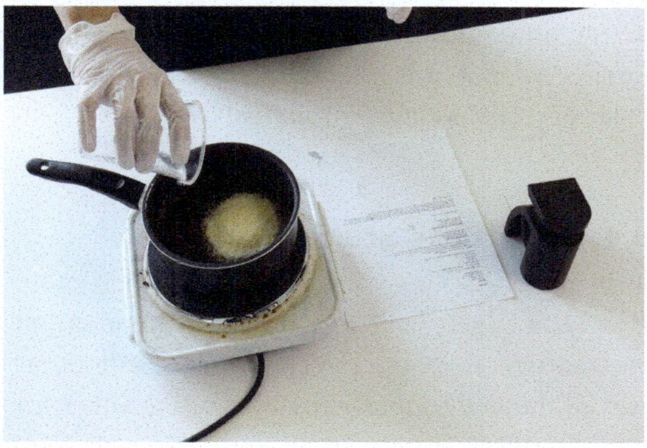

Fig. 4.45 Finally 12 g of glycerol are added. The volume of each part can be adjusted as long as the ratio 20:4:1 remains roughly the same. Less glycerol will create a more brittle but harder material, more glycerol will make a flexible and softer sheet (Roman and Pantazis, Bioplastic Tutorial 2012)

Heating (Fig. 4.46).

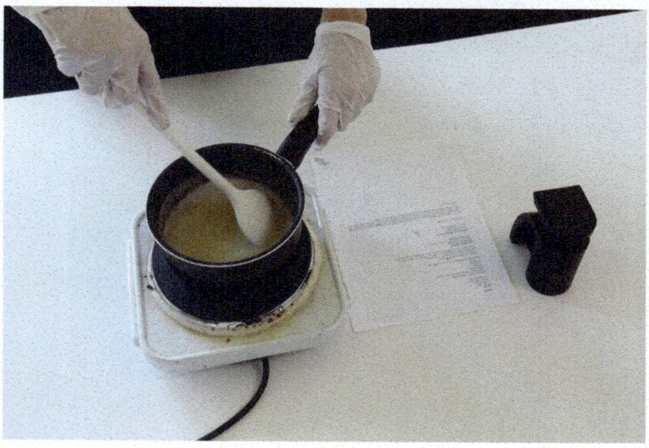

Fig. 4.46 The mixture has to be stirred until no clumps remain and it is as dispersed as possible. The mixture is then heated, while continuously stirring, up to 95 °C or until it starts to froth. Once it has reached its peak temperature no more heat should be added and excess froth removed with a spoon (Roman and Pantazis, Bioplastic Tutorial 2012)

Molding (Fig. 4.47).

Fig. 4.47 Once the solution is ready and no more clumps remain it can be poured into the previously prepared mold. Aluminum foil can be placed underneath and greasing the mold with vegetable oil will help releasing the plastic after it has hardened. The liquid should be equally spread across the surface (Roman and Pantazis, Bioplastic Tutorial 2012)

Drying (Figs. 4.48, and 4.49).

Fig. 4.48 The amount of time it takes the material to dry strongly depends on the thickness of the final product as well as the temperature and humidity in the room. It might take several hours until the sheet has fully dried, a toothpick can help to check its state (Roman and Pantazis, Bioplastic Tutorial 2012)

Fig. 4.49 Once it is ready the sheet can be carefully removed from the mold using a knife or scalpel and then be further processed (Roman and Pantazis, Bioplastic Tutorial 2012)

4.4.3.3 Alternative Recipes

Casein-based bioplastic from milk

1 L Milk/4 Tablespoons vinegar.

The milk has to be slowly warmed in a pan or pot. The vinegar is then added while the milk is continuously stirred until solid clumps begin to form. The liquid is then poured through a strainer. The remaining clumps can be scooped out and left on a flat surface covered with kitchen paper. Kitchen paper is also used to press out excess moisture. The material will need several days to dry.

Casein-based bioplastic from cream and lemon juice

1/2 Cup heavy cream/Lemon juice.

The cream is mixed with three tablespoons of lemon juice in a pan or pot. The mixture is then slowly heated while continuously being stirred. Gradually, more lemon juice is added while stirring further. The mixture will eventually thicken and make a gel-like consistence. After straining the mixture the solid cream by-products can be collected and washed. They can then be molded or shaped into various forms.

Starch-based bioplastic

700 ml Water/100 ml Vinegar/50 ml Glycerin/150 ml Cornstarch.

The water, vinegar, and glycerin are poured into a pan or pot and mixed. Then cornstarch is added and the solution thoroughly stirred, while slowly adding heat, until it has completely dissolved. After continuously mixing and heating the liquid for about 10 min it begins to thicken and turns gel-like. The heat can then be turned off but the gel should be stirred for another 1–2 min before pouring into a mold or on a flat surface. The material needs between 12 and 24 h to dry, heavily depending on its thickness. Serious shrinkage might occur. The amounts and volumes can be adjusted in order to create materials with varying properties. More glycerin will result in a harder plastic, more starch makes for a denser and less viscous material. An eight parts water, 1 part vinegar, 1 part glycerin, and 1.5 parts starch mixture might be best for filling molds and easier to work with. However, starch does not make for a very sturdy plastic. It can be used to create thin, flexible films but is generally too weak to make solid objects like cups or utensils.

4.4.4 Case Studies

Bioplastic has been used in a number of exploratory student projects and workshops as part of the *materiability* research at the Chair for CAAD, ETH Zurich, both for their structural capabilities and aesthetic appearance.

4.4.4.1 Phototropia (2012)

During the MAS CAAD project *Phototropia* (*2012*) a group of four students extensively looked into different ways in producing bioplastic and in

Table 4.2 Table comparing various properties of different types of bioplastic (Kretzer, Bioplastic Comparison 2012)

	Curing	Molding resolution	Stability	Color	Surface
Sugar + natural fibers	Fast (<1 h), only in fridge	Good, precise	Good due to integrated fibers	Transparent but fibers are visible	Sticky
Agar	Good (18 h)	Good	Good	Opaque	Rough
Starch	Long (>24 h)	Good	Brittle	Yellow, translucent	Soft, smooth
Lactose	Very long (>48 h)	Rough	Very brittle	White/yellow, opaque	Rough
Gelatine	Very long (>48 h)	Very good	Very stable	Yellow, transparent	Very smooth

modifying both the basic recipe and the fabrication procedure in order to enhance its properties. During the first tests the students experimented with varying recipes and additives. They produced small bioplastic samples based on starch, sugar, agar, lactose, protein, as well as gelatin, whereas the latter proved to be the most applicable, and reinforced the material with fibers, textiles, and wires or molding it into three-dimensionally structural shapes. The following Table 4.2 provides a brief overview of different bioplastic recipes and their assorted properties.

Besides trying to improve strength, rigidity, brittleness, and longevity of the bioplastic components a major interest was in reducing the amount of necessary material, and hence reducing their weight. Using an air-compressor the students developed a technique to aerate the bioplastic fluid, which was based on a gelatine-glycerine mixture, and thus create extremely lightweight but stable structures (Fig. 4.50).

Fig. 4.50 Close-up photograph of aerated bioplastic structures (Roman and Pantazis, Bioplastics 2012)

Even though the aerated three-dimensional structures were the visually and aesthetically most interesting results, the time and effort it took to produce them became impracticable when considering the dimensions of the desired installation. In order to simplify the fabrication process the students thus developed a method to create thin aerated sheets that ought to become part of the membrane of the installation. In parallel the students produced hollow bioplastic struts that were later supporting the installation and house the electronic connections (Fig. 4.51).

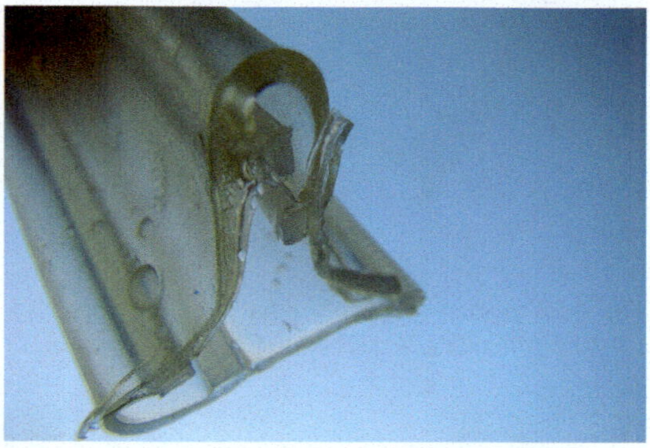

Fig. 4.51 Close-up photograph of bioplastic strut, which were holding the installation (Roman and Pantazis, Bioplastics 2012)

4.4.4.2 Bioplastic Formfinding Workshop (2012)

The *Bioplastic Formfinding Workshop* took place at the *Architecture Biennale PLAN12* in Cologne on September 22nd, 2012 and was held by Mauricio Rodriguez and Sachs at the exhibition and studio space of the architecture office 'Superartitecture' in Cologne.

During the 5-h workshop a group of students from different universities in Cologne as well as PLAN12 visitors produced a variety of bioplastic shapes and forms. The bioplastic was based on a gelatin-glycerin recipe, similar to the previous tutorial, and enhanced with various additional materials, including colors, fibers, or other substitutes. The aim of the workshop was mainly in providing a playful ground for do-it-yourself exploration of an interesting material and less in evaluating the potential of bioplastic as an applicable architectural element. The workshop encouraged the participants to experiment and self-make beyond the course of the workshop and initiated a cross-disciplinary exchange and discussion on the concept of open source and DIY. Throughout the workshop various prototypes were produced (Figs. 4.52, 4.53, 4.54 and 4.55), which the participants took home after they had fully dried.

Fig. 4.52 Variety of shapes produced during the Bioplastic Formfinding Workshop (Sachs, Bioplastics 2012)

Fig. 4.53 Variety of shapes produced during the Bioplastic Formfinding Workshop (Sachs, Bioplastics 2012)

Fig. 4.54 Variety of shapes produced during the Bioplastic Formfinding Workshop (Sachs, Bioplastics 2012)

4.4.5 Applications

Due to its organic compounds bioplastic obviously proposes a valuable alternative to fossil fuel-based plastic. In general two different types and usages need to be considered.

- Items that are mostly intended for packaging or single use applications and that quickly become obsolete. These should be produced to easily degrade, either in industrial composting systems or in the open air or water.
- Elements or products that are ought to stay permanent and where durability and robustness is essential. These need to exhibit equal or comparable properties like their oil-based equivalents but remain largely recyclable.

Despite the various advantages that bioplastics might have like the significant reduction of CO_2 emission, decreased dependency on fossil fuels, and less toxic waste they obviously still have to come a long way until they will be able to seriously compete their petrochemical rivals. Mostly, the two to three times higher cost in production but also the fear of loosing land for the growth of food or accelerated rate of deforestation hinder its economic development. Similar concerns exist over its impact on water supply and soil erosion.

Since the building and construction industry is among the largest consumers of plastics, the potential of less pollutant plastic alternatives is fairly obvious. However, the biodegradability of bioplastic poses a major problem and currently results in applications mainly reduced to the interior or of temporary character.

In 2011, Gernot Riether and a team of students at the Georgia Institute of Technology developed the 'AIA pavilion,' which was exhibited in New Orleans, USA. The 18 m^2 pavilion consisted of 320 different modules made from glycol modified polyethylene terephthalate (PET-G), a clear amorphous thermoplastic produced from sugar cane. The material can be injection molded or sheet extruded and is fully recyclable. The cells of the pavilion were prefabricated and partially preassembled in order to reduce setup time to only 2 days on site (Riether 2011).

In 2012, the Dutch architecture firm DUS began the development of a large-scale 3D printer that can produce objects using PLA. While the material has already become a strong competitor to the commonly used acrylonitrile butadiene styrene (ABS) in 3D printing, the concept of the 'Kamermaker' is to print entire habitable rooms (Kamer Maker).

In 2013, the institute of Building Structures and Structural Design (ITKE) at the University of Stuttgart, developed a new thermoplastic material based on over 90 % renewable resources. The bioplastic material called 'Arboblend' is made by combining different polymers such as lignin with natural fibers and is specifically designated for exterior applications in the construction industry. The material was physically demonstrated on a 140 m^2 freeform facade, the 'ArboSkin' (ITKE University of Stuttgart 2013).

4.5 Dye-Sensitized Solar Cells

A solar cell is a photonic device, which is able to harvest the energy of light with specific wavelengths and turn it into electricity. The quality of a photovoltaic (PV) cell is depending on its energy conversion efficiency, which is the percentage of solar energy shining on the cell that gets converted into electricity. Current commercially available silicon PV cells range in between 12 and 15 %. In June 2013, Sharp Corporation set the record for the World's most powerful research solar cell, based on a concentrator triple-junction compound, with an efficiency of 44.4 % (Sharp Press Releases 2013). The record was topped in 2014 by Fraunhofer, who produced a multi-junction solar cell converting 46 % of solar light into electrical energy (Fraunhofer 2014). Figure 4.55 gives an overview of different PV technologies and their development and improvement from 1976 until today. A major challenge for traditional silicon-based technologies remains the extremely high cost in processing the material. Intense research is hence focusing on the development of a low-cost, highly efficient solar cell that remains easy to manufacture.

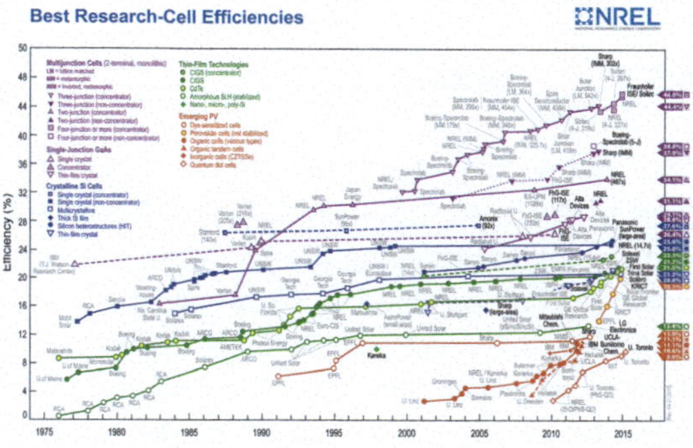

Fig. 4.55 Overview of solar cell conversion efficiency improvement from 1976 until 2015 for different technologies (this plot is courtesy of the National Renewable Energy Laboratory, Golden, CO, 2015)

Among the different technologies for solar energy creation, represent dye-sensitized solar cells (DSC) a very promising, emerging technology, and viable economical alternative. A dye-sensitized solar cell (Fig. 4.56) is a nanostructured photoelectrochemical device, consisting of a highly porous film of titanium dioxide (TiO_2) particles, coated with a monolayer of dye that is sensitive to visible light. Similar to the chlorophyll in plants, which produces chemical energy, the dye has the role of absorbing solar energy and transforming it into electricity. The effect is hence often dubbed synthetic photosynthesis (Junghänel 2007, p. 20). Dye-sensitized solar cells are made from relatively cheap and widely available materials through simple and inexpensive processes. They can be transparent or differently colored, are lightweight, and flexible and exhibit high power conversion efficiency even under low, diffused, or indirect light conditions.

Fig. 4.56 Dye-sensitized solar cells are extremely thin and flexible and hence interesting for lightweight applications (Kretzer, Solar Cell 2013a, b)

4.5.1 History

The photovoltaic effect was first discovered by French Physicist Alexandre-Edmond Becquerel in 1839 while observing the production of current between two platinum electrodes immersed in a halide salt electrolyte solution and exposed to light, an experiment that was strongly motivated by recent developments in photography (Grätzel 2001, p. 338). In 1883, the German photochemist Hermann Vogel discovered that by adding small amounts of an aniline-based dye to the photographic silver halides emulsion, which at that time was only sensitive to blue, violet, and ultraviolet light, its sensitivity could be enhanced toward mid-spectrum and red light (Grätzel 2003, p. 147). Four-years later, James Moser carried the concept of dye-enhancement over from photography to photoelectrochemical cells by using the dye **Erythrosine**.

A direct link in between the two phenomena, photography and photoelectrochemistry, was presented by S. Namba and Y. Hishiki at the 1964 *International Conference on Photosensitization of Solids*, who used dye-sensitization on metal oxides (ZnO) (Wenger 2010, p. 15). In order to understand and simulate organic photosynthesis, the phenomenon was further explored at the University of California, Berkley in the late 1960s. In 1972, for the first time photons could be converted into electrical energy by charge injection of excited dye molecules into a wide bandgap semiconductor (Junghänel 2007, p. 6). While research continued the instability and poor efficiency of the devices (around 1 %) remained a major problem. By introducing sensitized nanoporous Titanium Dioxide (TiO_2) electrodes, which dramatically increased the light harvesting surface area, a conversion efficiency of 7.1 % could be achieved in 1991 by Brian O'Regan and Michael Grätzel (Grätzel 2003, p. 147). Since then research has propelled and as of mid 2013 researchers around the group of Michael Grätzel at EPFL successfully created a state solid dye-sensitized solar cell version that raised the cell efficiency to 15 % (Burschka et al. 2013, p. 318). During the same time, Yun Hang Hu, scientist at Michigan Technological University, developed a 3D graphene counter electrode in order to replace the relative expensive platinum parts without loosing efficiency (Wang et al. 2013, p. 9210). In August 2013 Researchers at Stockholm's KTH Royal Institute of Technology presented a way to increase solar cell efficiency by 20 % through the introduction of a quasi-liquid polymer-based electrolyte (Achari et al. 2013, p. 17424) while just 1 month earlier scientists from the University of Basel developed a copper-based dye-sensitized solar cell where the relatively expensive element iodine had been replaced with the 50 times more abundant element cobalt (Bozic-Weber 2013, p. 7222).

4.5.2 Material Structure and Operation

A typical dye-sensitized solar cell consists of five parts (Fig. 4.57). The top part is usually a glass substrate (1) coated with a transparent conducting oxide (TCO), often a fluorine-doped tin dioxide (SnO_2:F) or indium tin oxide (ITO). In order to produce flexible dye-sensitized solar cells ITO-coated plastics have successfully been used instead of glass (Miettunen et al. 2013, p. 106). The backside of the conductive plate is covered with a thin layer, around 10 μm, of nanocrystalline TiO_2 that acts as the semiconductor (2). The TiO_2 forms into a highly porous structure with a surface area that is over a thousand times larger than the projected surface area.

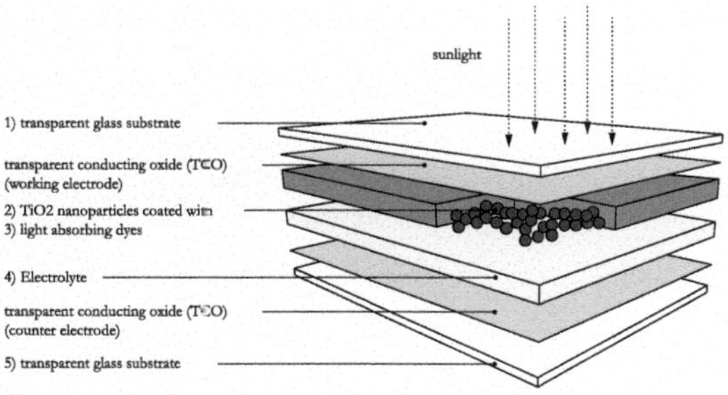

Fig. 4.57 Various layers of the nanostructured electrochemical dye-sensitized solar cell (adapted with permission from Kalyanasundaram and Grätzel 2009, Fig. 2)

Hence, a 1 cm^2 film has an actual surface area of 1000 cm^2 (Hara and Arakawa 2003, p. 666). The plate is then soaked in a dye (3), which forms a monolayer on the surface of the TiO_2.

Since the nanoporous TiO_2 has a much larger surface area than a single-crystal layer would have, the amount of absorbed dye is dramatically increased and more solar energy can be captured. Two types of dye are generally applied, metal–organic complexes and metal-free organic dyes (Ma and Zhan 2013, p. 437).

The most common and efficient dye-sensitized solar cells are currently based on ruthenium (Ru) dyes, however, the limited availability of ruthenium complexes will eventually become a problem for large-scale production. A thin layer of iodide electrolyte (4) is then spread along the TiO_2 and permeates its porous structure to form a **redox mediator**. The complete assembly is sandwiched together with the counter electrode (5), typically a platinum-coated TCO glass substrate. The two glass plates are finally sealed to prevent leakage of the electrolyte and evaporation of the solvent.

When the finished cell is exposed to sunlight (hv) the dye photosensitizer (S) on the TiO_2 surface absorbs incident photons of sufficient energy. These photons stimulate the dye molecules from their ground state (S) to their excited state (S^*).

$$S + hv \rightarrow S^*$$

Next the excited electrons get injected into the conduction band of the TiO_2 semiconductor, which in turn leads to the oxidation of the dye (S^+).

$$S^* \rightarrow S^+ + e^-(TiO_2)$$

The electrons are then transported throughout the TiO_2 layer toward the back contact (TCO) and finally reach the counter electrode through the external circuit. By reducing the oxidized S^+ within the electrolyte it arrives back at its ground state (S) while the iodide redox mediator (I^-) gets oxidized to triiodide ($I3^-$).

$$S^+ + e^- \rightarrow S$$

Finally, the loop is closed when the oxidized redox mediator ($I3^-$) restores its missing electron (I^-) by diffusing toward the counter electrode.

$$I_3^- + 2e^- \rightarrow 3I^-$$

4.5.3 Fabrication

The fabrication of a dye-sensitized solar cell is relatively simple and do-it-yourself kits for school or class experiments are widely available (Solaronix). The TiO_2 powder is a commercial product, which is usually applied on the TCO substrate using the **doctor blade** technique or via screen-printing. In either case the films posses a thickness between 5 and 15 μm and porosity of 60–70 %. After the TiO_2 film has been prepared it is immersed into the dye-solution and then stored at room temperature for 12–18 h. This process results in intense coloration of the film, which is subsequently washed with alcohol to remove nonabsorbed dye residue. The counter electrode is in most cases made by sputtering Platinum on a TCO substrate. The electrolyte solution is then dropped onto the surface of the TiO_2 electrode and the two electrodes sandwiched and sealed together using a copolymer of ethylene or acrylic acid.

While most solar cell kits focus on creating rigid cells encased in glass substrates, the following tutorial produces a flexible dye-sensitized solar cell using a thin titanium sheet. While, this makes the fabrication process slightly more difficult it vastly enhances the scope of application and reduces the weight of the resulting system.

4.5.3.1 Materials and Tools

See Fig. 4.58.

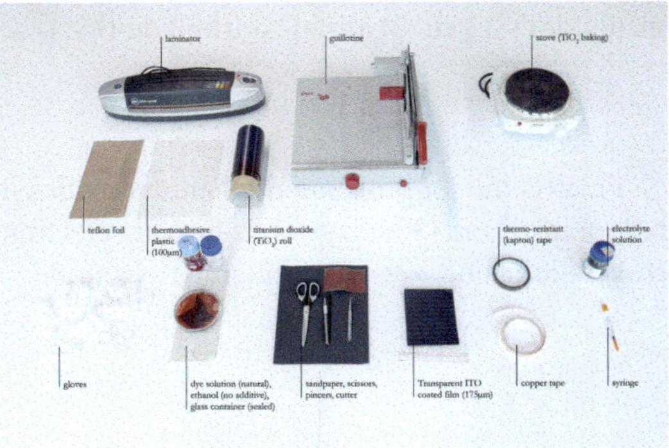

Fig. 4.58 Laminator/guillotine (paper cutting machine)/hot-plate (with heat-control to heat up to 500 °C)/Teflon foil, thermo-adhesive PET plastic foil (ca. 100 µm)/Titanium dioxide (TiO_2) coated titanium sheet/thermo-resistant tape (Kapton 12 mm × 33 m)/dye-solution (ruthenium-based dyes are most efficient but organic dyes such as blueberry or rhododendron can also be used)/ethanol/liquid electrolyte/glass container/petri dish or container with lid/transparent ITO-coated film (175 µm)/conductive tape (copper or aluminum)/sandpaper/tweezers, pincers, scissors, cutter/syringe/kitchen paper/gloves (Roman and Pantazis, Dye-Sensitized Solar Cell Tutorial 2012)

4.5.3.2 Steps

Plastic container (Figs. 4.59 and 4.60).

Fig. 4.59 The plastic container will house the DSC. First the desired shape of the cell has to be cut out twice of the adhesive PET plastic foil. The second piece needs a small opening in the center for the electronic connection (Roman and Pantazis, Dye-Sensitized Solar Cell Tutorial 2012)

Fig. 4.60 The two layers are then laminated together using a Teflon sheet to avoid melting of the foils (Roman and Pantazis, Dye-Sensitized Solar Cell Tutorial 2012)

Transparent front electrode (Fig. 4.61).

Fig. 4.61 The front electrode is prepared from the conductive plastic foil and needs to be the exact same shape as the later DSC. To one edge of the foil conductive tape has to be added, which will become the positive connection and ease the transfer of electrons (Roman and Pantazis, Dye-Sensitized Solar Cell Tutorial 2012)

Flexible TiO_2 foil (Figs. 4.62, 4.63 and 4.64).

Fig. 4.62 Using the guillotine and scissors the shape of the DSC now has to be cut out from the metallic TiO_2 foil. The edges need to be sharp and clean to avoid short-circuits (Roman and Pantazis, Dye-Sensitized Solar Cell Tutorial 2012)

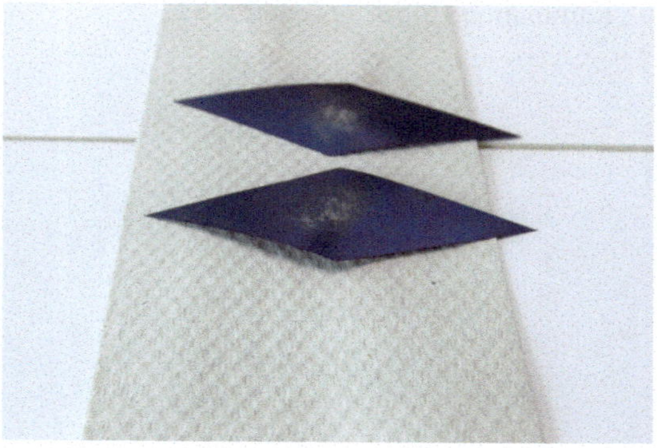

Fig. 4.63 In order to fit into the previously prepared plastic containers, the pieces should be slightly smaller. From the container one edge has to be removed, which will be used to inject the electrolyte (Roman and Pantazis, Dye-Sensitized Solar Cell Tutorial 2012)

Fig. 4.64 As TiO_2 is highly porous and easily absorbs humidity and dust from the air. Therefore the sheet has to be baked at 500 °C for at least 1 h using the hot plate, until the sheet turns *bright white* (Roman and Pantazis, Dye-Sensitized Solar Cell Tutorial 2012)

Dyeing (Fig. 4.65).

Fig. 4.65 After baking the TiO_2 sheet has to be soaked in the dye solution for a few hours, preferably over night. During this time the dye should be stored in a dark area and remain covered with a lid. After dyeing the piece needs to be dried thoroughly and the colored side then moistened with a bit of ethanol (Roman and Pantazis, Dye-Sensitized Solar Cell Tutorial 2012)

Lamination (Figs. 4.66 and 4.67).

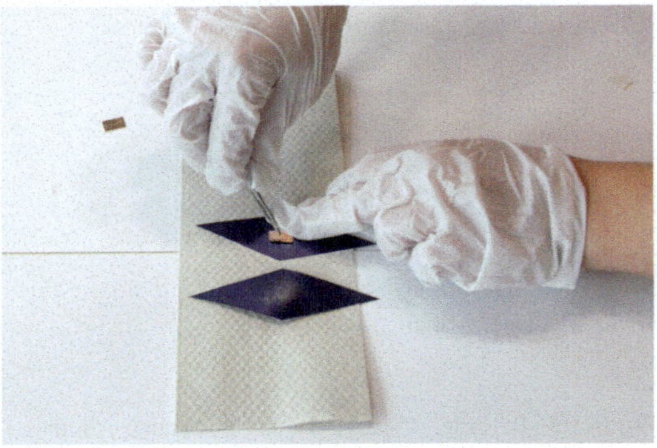

Fig. 4.66 The previously prepared transparent front electrode is now fixed on top of the dyed TiO_2 foil (Roman and Pantazis, Dye-Sensitized Solar Cell Tutorial 2012)

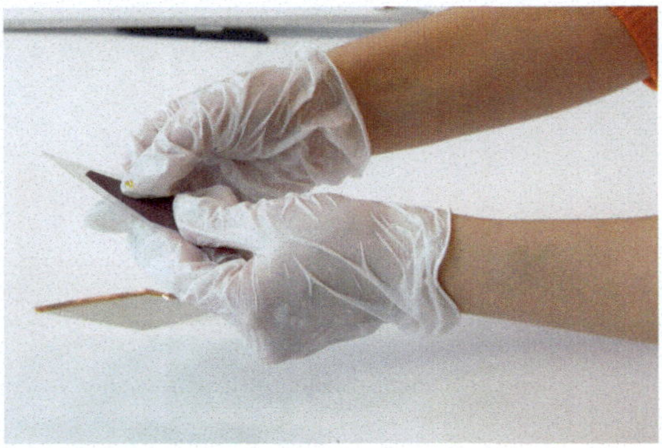

Fig. 4.67 Both pieces are then placed inside the plastic container and glued together using the laminator. One edge needs to be *left* open for electrolyte injection (Roman and Pantazis, Dye-Sensitized Solar Cell Tutorial 2012)

Electrolyte (Fig. 4.68).

Fig. 4.68 Now the electrolyte can be filled into the container using a syringe and needle. Finally the small injection hole can be closed with a piece of thermo-resistant tape (Roman and Pantazis, Dye-Sensitized Solar Cell Tutorial 2012)

Current (Fig. 4.69).

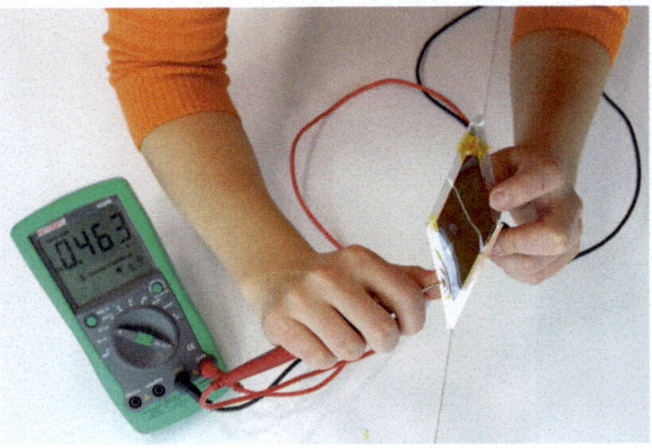

Fig. 4.69 Using a multi-meter the dye-sensitized solar cell can be tested for functionality. The output voltage is independent from the size of the component but the current changes in relation to the amount of incoming sunlight (Roman and Pantazis, Dye-Sensitized Solar Cell Tutorial 2012)

4.5.4 Case Studies

The research into dye-sensitized solar cells emerged during the preparation for the MAS CAAD module *Phototropia* in 2012. The main idea was to combine a number of (self)-active materials into an installation that would not only respond to external input but also create all its required energy by itself.

4.5.4.1 Phototropia (2012)

The project *Phototropia* was realized within 5 weeks in spring 2012 by 19 students at the Chair for CAAD, ETH Zurich, who worked in four different teams. The main focus of the project was to prove that a completely autonomous, self-sufficient active installation could be produced entirely from self-made smart materials. Each team worked on a different material, electroactive polymers, electroluminescent displays, bioplastics, and dye-sensitized solar cells (Kretzer 2013a, b, pp. 211–212).

During the first week of the course, the students focused entirely on understanding and reproducing their chosen material. By modifying certain steps during the fabrication process or adjusting material quantities they were not only able to comprehend how the materials were

Table 4.3 Table displaying DSC size/shape and efficiency ratio, with best conversion highlighted (Pantazis and 2011/12 MAS CAAD class, DSC efficiency comparison 2012)

					Shadow 45°			Sun 45°		
No	Shape	Width (cm)	Height (cm)	Area (cm^2)	Voltage (V)	Current (mA)	I/A (mA/cm^2)	Voltage (V)	Current (mA)	I/A (mA/cm^2)
1	Rectangular	5.5	5.5	30.3	0.6	2.0	0.066	0.7	6.0	0.198
2	Rectangular	3.5	3.5	12.3	0.5	0.6	0.049	0.7	10.0	**0.816**
3	Rectangular	27.0	3.0	81.0	0.6	6.0	0.074	0.7	40.0	0.494
4	Triangular	4.0	3.5	7.0	*broken*	*broken*		0.23	*broken*	
5	Rectangular	12.0	8.5	102.0	0.6	7.3	0.072	0.7	23.0	0.225
6	Rectangular	12.0	8.5	102.0	0.6	7.6	0.075	0.7	30.0	0.294
7	Triangular	3.5	2.5	4.4	0.5	0.1	0.023	0.6	1.5	0.343

structured and how they functioned but also to change their behavior according to desired design outcomes. The team working on dye-sensitized solar cells began by empirically identifying the relationship between size and shape of the device and the generated current and voltage in different sun exposure situations (Table 4.3).

The produced voltage remained roughly the same among all samples and did not seem to be related to neither size nor shape. The current however differed strongly among the various designs with the larger shapes producing considerably more than the smaller ones. Interestingly even though the larger cells produced a greater amount of current and exhibited better conversion efficiency under indirect illumination, a small square shape of 3.5 cm × 3.5 cm displayed the best ratio under direct sun exposure. The difference between indirect and direct solar radiation proved to be much more significant than initially estimated.

Even though the empirical tests did, due to their experimental nature, not exactly follow the mathematical estimations, the conclusion was to produce a number of shapes, which would be serially connected in clusters in order to reach a higher voltage and then connect the respective clusters in parallel to raise the amount of current.

In addition to these empirical studies, various designs of the overall installation were proposed and discussed. The goal became to combine the dye-sensitized solar cells and electroluminescent elements into a single membrane that should hover above ground and become kinetically activated through electroactive polymers that were to move the bioplastic struts. A number of rigid origami tessellations were then examined for their feasibility as kinetic folding structures (Fig. 4.70).

Fig. 4.70 Rendering of proposed origami tessellation and design (Pantazis, Rendering 2012)

Once the overall layout and shape was defined different arrangements and patterns of the various elements were tested (Fig. 4.71). The ratio of dye-sensitized solar cells to electroactive polymers and electroluminescent displays had to be roughly two-third to one-third.

Resulting from the particular structure of the origami tessellation the individual components became regular triangular pieces with their short edges being 15 cm wide. The size of the incorporated dye-sensitized solar cells was 12.5 cm 12.5 cm (Fig. 4.72). The dye-sensitized solar cells were facing upwards while the electroluminescent displays were oriented toward the ground.

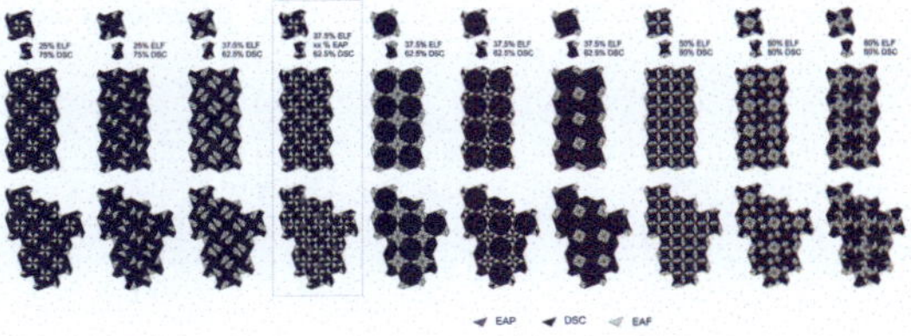

Fig. 4.71 Different patterns emerging from the combination of EL and DSC (Alvarez-Marin, Patterns of Phototropia installation 2012)

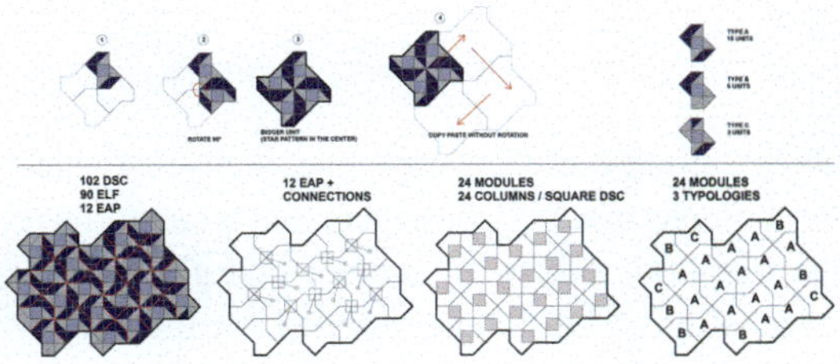

Fig. 4.72 Final pattern, variations across the installation and location of various elements (Alvarez-Marin, Patterns of Phototropia installation 2012)

The final design was supposed to consist of 102 custom made dye-sensitized solar cells, 90 electroluminescent displays, and 12 electroactive polymers. Since the required power was more than the custom made solar cells could generate 24 industrially fabricated solar cells had to be integrated. Each custom made element was able to produce 0.5 V, the industrial component released 3 V. Always twelve self-made dye-sensitized solar cells were then connected in series with one industrial cell, which resulted in a total output per cluster of 9 V.

The generated energy was then stored in batteries below a base platform and when these were charged sufficiently, distributed via microcontrollers to the corresponding elements. An integrated infrared sensor ensured that the installation would only become active when visitors were present.

The installation was built and displayed at the Chair for CAAD, ETH Zurich (Figs. 4.73, 4.74 and 4.75). The final setup was 150 cm wide, 180 cm deep, and about 80 cm in height. Since the gelatin-based bioplastic, which constituted the structural parts of the project, however remained very sensitive to changes in temperature and especially humidity, the installation slowly collapsed after a few weeks of exhibition.

Fig. 4.73 Final installation as exhibited at the Chair for CAAD, ETH Zurich in April 2012 (Pantazis and Kretzer, Phototropia 2012)

Fig. 4.74 Final installation as exhibited at the Chair for CAAD, ETH Zurich in April 2012 (Pantazis and Kretzer, Phototropia 2012)

Fig. 4.75 Final installation as exhibited at the Chair for CAAD, ETH Zurich in April 2012 (Pantazis and Kretzer, Phototropia 2012)

4.5.5 Applications

In order to become commercially available and applied in a larger scale, dye-sensitized solar cells still have to master a number of challenges. While conversion efficiency is continuously improved and already comparable with other photovoltaic technologies, a major problem remains the cell's low stability. To be market competitive, the cells need to be intrinsically and extrinsically stable toward changing temperatures, humidity and illumination for about 20 years with a loss of no more than 10–20 %, which is what other common manufacturers guarantee. Moreover, the low scalability, mostly due to the liquid electrolyte, of the cells poses a large issue when thinking about architectural applications. However, motivated by constant improvements, scientists are convinced to solve these challenges soon and a number of companies, including DyeSol in Australia, Solaronix in Switzerland and G24innovations in the UK, are already promising near-future market introduction of DSC technology.

Due to their flexibility, transparency, and lightweight dye-sensitized solar cells offer great potential for architectural usage, reaching from transparent or colored energy-producing skylights or windows to flexible curtains or membranes that could be used in temporary structures or emergency shelters. Their ability to work in low-light conditions makes them also particularly interesting for indoor usage or areas that are not constantly exposed to direct sunlight.

The EPFL Swiss Tech Convention Center in Lausanne by local architects Richter-Dahl Rocha and Associates is as of early 2015 the World's first prototype applying large-scale dye-sensitized solar cell technology into a public building. The West facade includes 1500 red and orange shaded 35 × 50 cm solar power generating glass modules over a surface of 300 m^2. On top of the energy creating aspect (estimated at 8000 kWh per year, which of course only makes a small part of the total building consumption) these elements will reduce solar radiation and regulate the temperature inside the building, but most of all become aesthetically interesting fully integrated building components. The center was inaugurated in Spring 2014 (Barraud 2012).

4.6 Electroluminescent Displays

The two main ways of producing light are incandescence and luminescence. While in incandescence an electric current is driven through a conductor, which produces heat in resistance to the current and hence generates light, luminescence is generally referred to all forms of visible light due to other causes than temperature. Electroluminescence (EL) thus means the non-thermal creation of light emerging from the application of an electrical field to a substance (Fig. 4.76). The light-emitting substance is usually a doped luminescent crystal (phosphor), e.g., zinc sulfide (ZnS) (Barrow 1999, p. 94-1). Depending on the dopant, different colors can be achieved. The best results are achieved from a manganese-activated zinc sulfide (ZnS:Mn), which emits a yellow-orange color peaking at 585 nm (Ono 1995, p. 82). A greenish light can be produced using a copper dopant (ZnS:Cu) and a silver dopant (Zns:Ag) will result in a bright blue.

Fig. 4.76 Electroluminescent displays emit a perfectly homogeneous cold light and can be made as flexible foils (Kretzer, Electroluminescent Display 2013a, b)

The main types of electroluminescence are injection and high-field electroluminescence.

- Injection electroluminescence is the basic principle behind light-emitting diodes (LED) technology. It results from a semiconductor p-n junction being biased in the forward direction and connected to a source of DC voltage. In this case, either excess holes are injected into the **n-region** or excess electrons into the **p-region**. Luminescence occurs when the electrons and holes recombine (Mauch 1996, p. 589).

- High-field electroluminescence represents the functioning principle behind electroluminescent displays and is produced by impact excitation of a luminescent center by high-energy electrons (Ono 1995, p. 1). Electroluminescent displays are a type of flat panel display produced by sandwiching a light-emitting substance between two conductors. They are extremely rugged, relatively insensitive to ambient temperature, exhibit rapid display response (<1 ms), require very little current, and emit a soft, perfectly homogeneous light, which is visible from great distance and essentially unlimited viewing angle. They are compact, lightweight, inexpensive in production, and can be made in any color. The light output of electroluminescent displays is very reliable with typically only 10 % loss after 10,000 h of operation and an overall lifetime of more than 100,000 h.

4.6.1 History

The phenomenon of electroluminescence was first discovered by Captain Henry Joseph Round in 1907, when running current through a silicon carbide (SiC) detector. The next reported observation was in 1923, when Russian scientist Oleg Lossev confirmed the occurrence of yellow light emission in SiC crystals. In the late 1920s, Bernhard Gudden and Robert Pohl conducted experiments in Germany with copper doped ZnS. However it was not until 1936 when Georges Destriau, an associate of Marie Curie, discovered high-field electroluminescence and hence officially coined the term.

During World War II, a serious amount of research was performed on luminescent crystals in combination with radar displays but only in the 1950s companies like GTE Sylvania focused on powder electroluminescent phosphors and started developing the technology for practical applications. Nataliya Vlasenko and A. Popkov successfully fabricated the first thin-film electroluminescent (TFEL) structures in 1958 while observing that luminance was much higher when doping zinc with manganese (ZnS:Mn) as opposed to previously used copper (ZnS:Cu). Despite continuous progress in electroluminescent technology the devices remained too unstable for commercial use, until research revived in the mid 1960s with a focus on display applications. In 1965 Sigmatron Corporation demonstrated a TFEL dot-matrix display and in 1968 Aron Vecht developed DC electroluminescent technology to be used in lamps and watches. In 1974, Toshio Inoguchi at Sharp Corporation introduced an AC TFEL approach, which became the first high-brightness long life electroluminescent display and kept Sharp at the leading edge for the following few decades (Hart et al. 1999).

In the 1980s, Christopher King developed advanced thin-film electroluminescent displays and launched Planar Systems in the US in 1983. Both companies, Sharp and Planar Systems, produced monochrome EL displays that were applied in early laptop computers since 1985. In 1993, based on research by William Barrow and his team at Planar, the production of full-color Active Matrix electroluminescent (AMEL) microdisplays began. In the 1990s, Xingwei Wu at Canadian company iFire developed thick-film dielectric electroluminescent (TDEL) technology. TDEL is based on inorganic electroluminescent technology and is brighter, more efficient, and more resistant to contamination during fabrication and to electrical failure than thin-film electroluminescence. Some TDEL use two layers of phosphors. The thick bottom layer is resistant to dielectric breakdown and can withstand a higher current to produce brighter light, while the upper layer consists of colored phosphors. Through this RGB electroluminescent devices can be created (Heikenfeld and

Steckl 2003, p. 21). Mid-2007 Planar presented a transparent and segmented electroluminescent (TASEL) display in custom forms and sizes with curved and moldable shapes. TASEL displays are based on the standard electroluminescent device structure but a transparent electrode replaces the metal rear electrode (Abileah et al. 2008). In 2012, Cho et al. showed a hybrid AC electroluminescent (HACEL) device containing a thin organic/inorganic composite film of quantum dots blended in a light-emitting polymer matrix. The device exhibits high brightness and is meant to overcome current limitations in color blending and tuning (Cho et al. 2012, p. 4545).

4.6.2 Material Structure and Operation

Figure 4.77 displays the basic device structure of a thin-film electroluminescent display. The supporting substrate is usually made of very low sodium glass (1), coated with a transparent conductor such as indium tin oxide (ITO). In order to produce flexible displays a transparent polyethylene or polycarbonate film can be used as base substrate. The ITO layer is between 120 and 350 nm in thickness (2). On top of the ITO a lower insulator (200 nm) is deposited. The light-emitting phosphor (4) is typically 200–1000 nm in thickness and sandwiched between two insulating films (3, 5) (Barrow 1999, p. 2). Due to the very high-field level (1.5 MV/cm) that is necessary to excite the phosphor layer to emit light, any defect in the thin-film stack could produce a short circuit and destroy the device. Hence, current-limiting layers (the insulators) on both sides of the phosphor are essential (King 1996, p. 1). The back electrode is deposited on top of the upper insulator and is generally 100–200 nm thick (6). In most electroluminescent tape configurations, power is provided through copper or aluminum conductors. In electroluminescent sheets or panels different conductive materials can be used, depending on the particular design requirements.

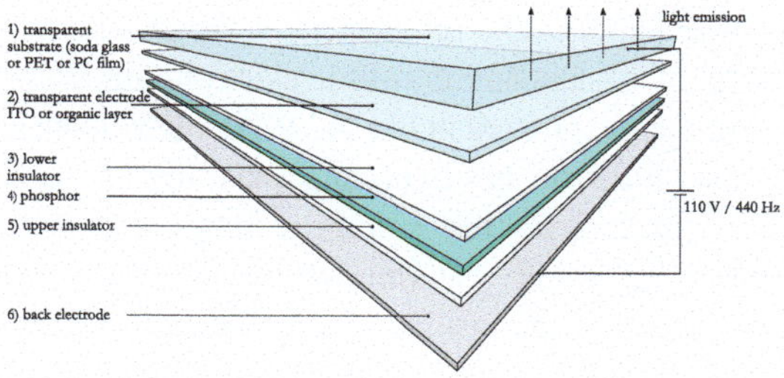

Fig. 4.77 Schematic diagram of basic thick-film electroluminescent structure (adapted from Enz Schematic diagram of basic thick-film electroluminescent structure 2010)

If an imaging matrix is desired, both electrodes are made of grids of parallel lines, with their orientations being orthogonal to each other. Using approximate circuitry the whole matrix can thus be scanned and voltage applied to any phosphor element that is located at the respective intersections (Fig. 4.78).

Fig. 4.78 Schematic of *dot*-matrix display. From *left* to *right*: rear electrode layer, light-emitting pigment, translucent front electrode (adapted from Brunner and Welker 2005, p. 325 with permission)

Depending on the size of the thin-film electroluminescent device, it can be powered from a small inverter, which transforms low DC voltage to the appropriate level of AC power, causing an electric field to occur in between the upper and lower electrode. During each half-cycle, electrons are emitted into the high-field region and get accelerated to ionize atoms of the crystal lattice. The resulting holes are trapped by luminescent centers. When the field is oriented into the opposite direction, due to the AC current, the electrons return to the luminescent centers, where they recombine with the holes to glow. The brightness of the display can be adjusted by controlling the frequency and voltage of the power supply. The luminance—voltage relationship of a standard thin-film electroluminescent display is expressed in Fig. 4.79. At low voltage it emits little or no light and acts like a pure capacitor. Once the voltage reaches a certain threshold value, determined by the dielectric properties of the insulating and phosphor layers, the brightness increases abruptly and steeply (six lumen/watt) until it reaches a saturated region, where efficiency goes down (three lumen/watt). The relative brightness of a thin-film electroluminescent display is roughly proportional to the driving frequency. Exceeding the maximum voltage and frequency will significantly reduce its lifetime.

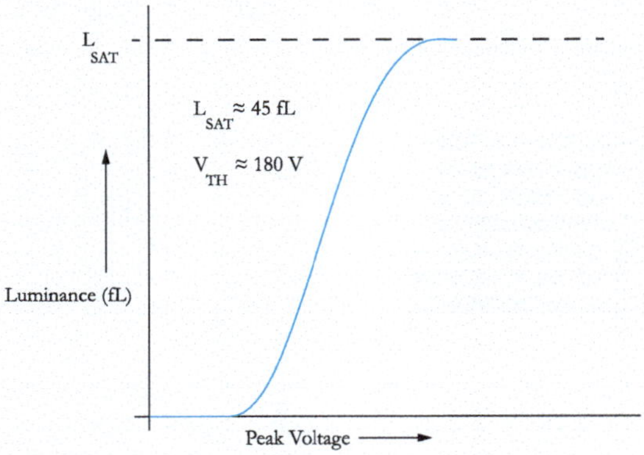

Fig. 4.79 Luminance versus voltage relationship of a TFEL device (adapted with permission from King 1996, p. 1, Fig. 2)

4.6.3 Fabrication

Thin-film electroluminescent displays are commercially produced using a various different deposition techniques. The electrode materials ITO and aluminum are often deposited by sputtering or **physical vapor deposition** (PVD). The insulator and phosphor layers are usually made by **chemical vapor deposition** (CVD) or PVD. But electroluminescent displays can also be fabricated using screen-printing techniques, which allow for reduction in material usage, less cost, and add speed and adaptability to the fabrication (Brunner and Welker 2005, p. 324). In screen-printing the display consists of four layers (Fig. 4.80):

- A transparent conductive layer, which is either an ITO-coated glass substrate or polyester film.
- A phosphor layer, which has to be evenly spread across the electrode.
- A dielectric insulator layer, whose thickness depends on the insulating requirements.
- A rear electrode that needs to match the inner edge of the bus bar and can either be from silver or carbon ink.

The electric field is created across the phosphor by applying an AC voltage from the transparent conductive to the silver layer (Sarik et al. 2011, p. 315).

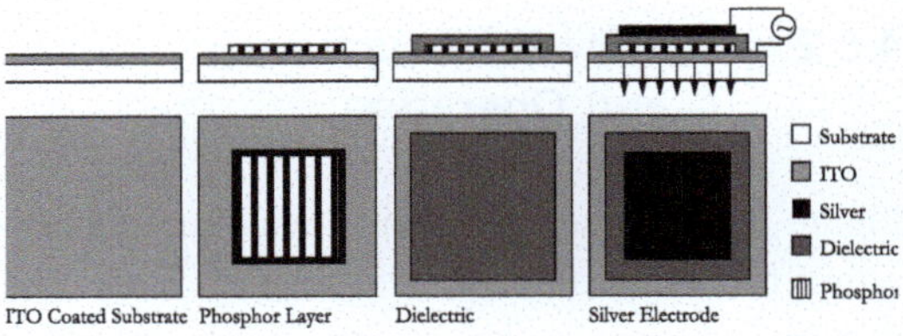

Fig. 4.80 Electroluminescent screen-printing steps (adapted with permission from Sarik et al. 2011, p. 315)

4.6.3.1 Materials and Tools

See Fig. 4.81.

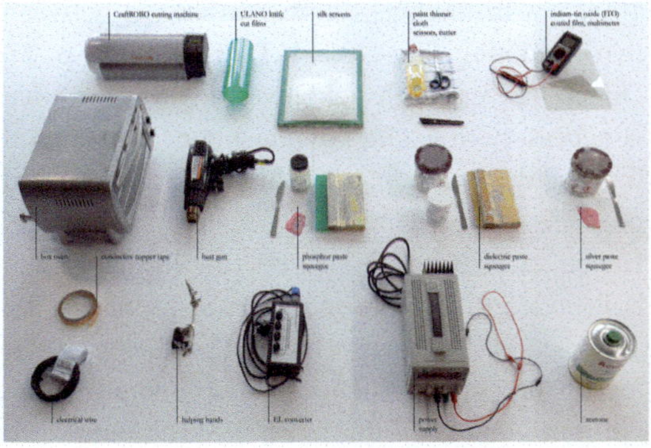

Fig. 4.81 Craftrobo or Silhouette cutting machine/Ulano knife cut films/silk screens/paint thinner, acetone/cloth, scissors, cutter, squeegee/ITO (indium tin oxide) coated foil/multi-meter, conductive tape, electrical wire/box oven, heat gun/phosphor paste, dielectric paste, silver paste/EL inverter, power supply/soldering iron, helping hands (Roman and Pantazis, Electroluminescence Tutorial 2012)

4.6.3.2 Steps

Electroluminescent pattern (Fig. 4.82).

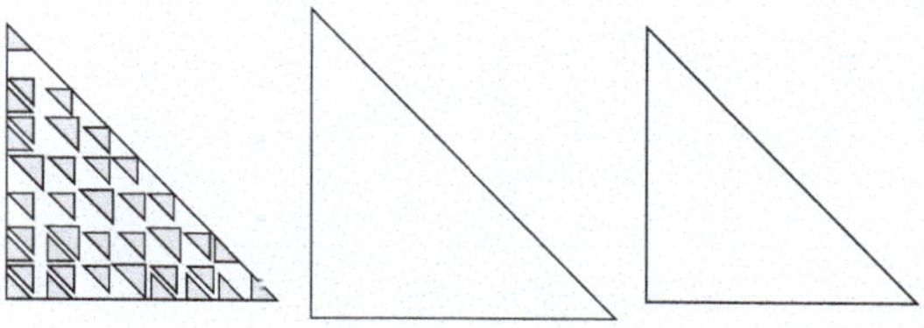

Fig. 4.82 The electroluminescent pattern design should be drawn on a computer using a graphical program and then applied to the knife cut film with a cutting machine. There are further ways to create stencils for screen-printing, like photo emulsion or paintable screen fillers; the proposed way however is easy, quick and convenient. It is important to have three different stencils, one for the phosphor, one for the dielectric, and one for the silver layer. The electroluminescent design only appears on the phosphor layer, which emits the light. The dielectric layer has to entirely cover the phosphor layer and should therefore extend 5 mm on each side. The silver layer needs to cover the phosphor layer but has to be smaller than the dielectric layer (Roman and Pantazis, Electroluminescence Tutorial 2012)

Film adhesion onto silkscreen (Figs. 4.83, 4.84, and 4.85).

Fig. 4.83 The cut-out parts need to be carefully removed from the knife cut film (Roman and Pantazis, Electroluminescence Tutorial 2012)

Fig. 4.84 Then the film has to be fixed with the rough side facing down onto the silkscreen. Paint thinner is used on the inner side of the silkscreen by firmly rubbing a soaked cloth onto the surface (Roman and Pantazis, Electroluminescence Tutorial 2012)

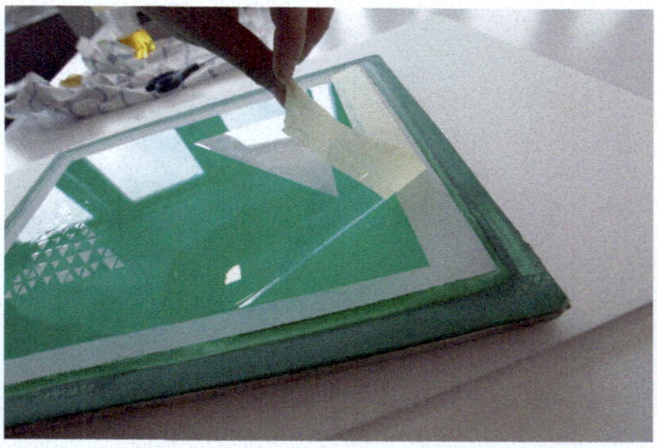

Fig. 4.85 Once the thinner has dissolved, the transparent part of the film can be removed (Roman and Pantazis, Electroluminescence Tutorial 2012)

Preparation of work area (Fig. 4.86).

Fig. 4.86 The perimeter of the silk screen as well as four times the outlines of the designed shape (*one* for phosphor, *twice* for dielectric, and *one* for silver) should be marked onto a clean, *white* surface (Roman and Pantazis, Electroluminescence Tutorial 2012)

Preparation of materials (Fig. 4.87).

Fig. 4.87 After checking the electrical conductivity of the ITO foil it needs to be fixed with its conductive side facing upwards onto the previously marked first position (phosphor) and the silk screen with the phosphor stencil placed *above* (Roman and Pantazis, Electroluminescence Tutorial 2012)

Phosphor printing (Fig. 4.88).

Fig. 4.88 A thin line of phosphor paste goes on the *top end* of the shape. Then, while pressing the squeegee down firmly the phosphor paste is printed onto the screen in one stroke. Instructions on screen-printing can be found online (Roman and Pantazis, Electroluminescence Tutorial 2012)

Drying and baking (Fig. 4.89).

Fig. 4.89 After using a heat gun to pre-dry the fresh print for about 2 min, the shape is baked inside a box oven for another 10 min at 130 °C. This has to be done carefully under constant supervision, as too much heat will burn the material (Roman and Pantazis, Electroluminescence Tutorial 2012)

Dielectric printing (Fig. 4.90).

Fig. 4.90 After placing the foil onto the dielectric position on the work area a thin layer of dielectric is printed. The foil is then dried and baked for 10 min at 130 °C. The step is repeated and a second layer printed on *top* of the first. These layers are the most important as they prevent the foil from shortening. Hence it is crucial to avoid holes in the dielectric layer. The occurrence of holes can be checked when holding the device against a source of light (this step should be repeated until all holes are covered) (Roman and Pantazis, Electroluminescence Tutorial 2012)

Silver printing (Fig. 4.91).

Fig. 4.91 The foil goes back onto the work area, this time onto the silver position. As in the previous steps, a thin layer of silver is printed and the foil then dried and baked for 10 min at 130 °C (Roman and Pantazis, Electroluminescence Tutorial 2012)

Adding connectors to active area (Figs. 4.92 and 4.93).

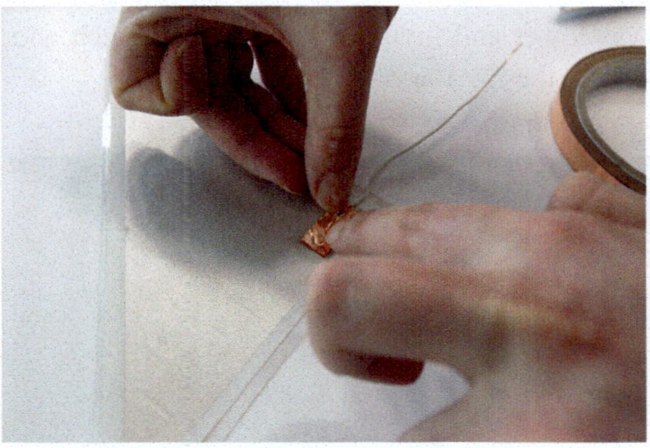

Fig. 4.92 A small strip of conductive (copper or aluminum) tape is attached on to the silver layer (Roman and Pantazis, Electroluminescence Tutorial 2012)

Fig. 4.93 The component will work better if the tape sits in an area with phosphor on the opposing side. Conductive tape is used again to fix an electrical wire on *top* of the previously attached tape (for better handling, soldering the wire onto the conductive tape makes the component more stable) (Roman and Pantazis, Electroluminescence Tutorial 2012)

Adding connectors to conductive film (Fig. 4.94).

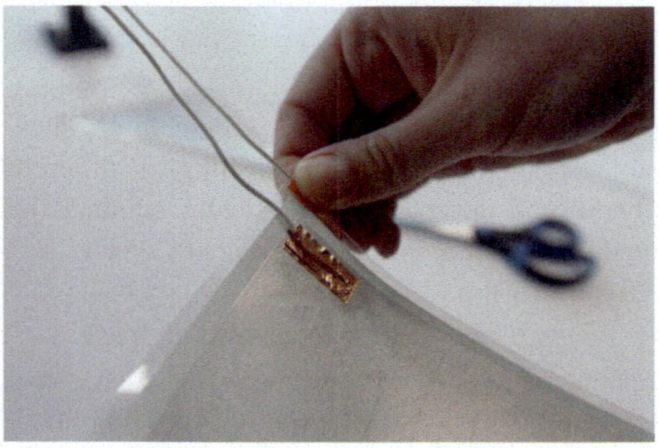

Fig. 4.94 As in step 9 conductive tape and wire need to be added to the ITO-coated edge of the foil. It is important not to touch the silver layer (Roman and Pantazis, Electroluminescence Tutorial 2012)

Power (Fig. 4.95).

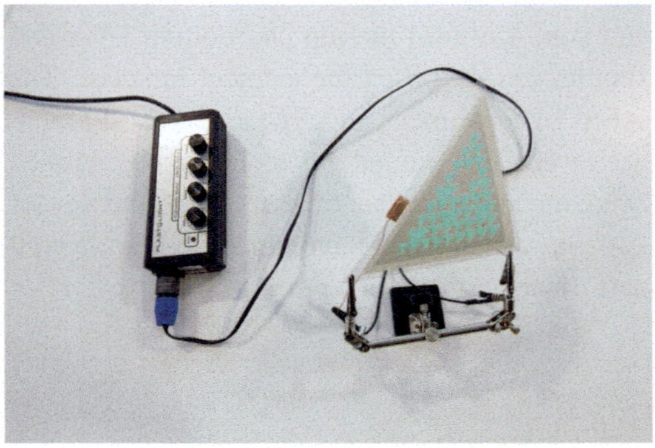

Fig. 4.95 Finally the component can be connected to the electroluminescent converter. As it's running on AC the polarity doesn't matter (Roman and Pantazis, Electroluminescence Tutorial 2012)

4.6.4 Case Studies

During the research at the Chair for CAAD, ETH Zurich, the properties of electroluminescence have been explored in a number of student classes and workshops. During the MAS CAAD module *Material Animation* in spring 2011 industrially produced electroluminescent sheets were used in order to create several kinetically animated light installations. The one-week workshop *Actuated Matter* focused on the self-making of such displays and took place in summer 2011 at the Zurich University of the Arts.

4.6.4.1 Material Animation (2011)

Material Animation was a series of kinetically moving light installations, made from laser-cut electroluminescent foils, which sensed location, number, and velocity of human occupants and responded through a multitude of wirelessly networked components to encourage further interaction.

The experiment was situated in three interconnected rooms, each reflecting a different theme and approach to physically animate the material in an architectural context (Kretzer and Glynn 2013, p. 35).

Vapor (Fig. 4.96) created a fluid space consisting of eight floating elements that were expanding and contracting in response to human interaction. Generative design processes helped to maximize the three-dimensional form, which was created by cutting and combining two A4 electroluminescent sheets. Each element was controlled and animated by two servo-driven pulleys, which were mounted to the back wall of the room. The servos allowed for the expansion and contraction of the top and bottom layer and simultaneously raised the element in space. The speed of movement and frequency of illumination were determined by a Java programmed behavior and the real time input of the sensor system, which was then tuned to the overall performance of the other spaces.

Fig. 4.96 Vapor consisted of a server-pulley-string system and was opening and closing according to user presence (Kretzer, Material Animation 2011)

Open Wires (Fig. 4.97) aimed to create an environment based on lighted, ephemeral, and unpredictable three-dimensional shapes. The system consisted of 31 electroluminescent strips that revolved and flickered in high speed. The strips were attached to the ceiling at two different heights. The first level was overhead, acting as a collective cloud system. The second position was lower and invited the visitors to touch and distort the ray trace.

Fig. 4.97 Open wires was made from thin electroluminescent strips, which flickered around visitors in high speed (Kretzer, Material Animation 2011)

Each element mainly consisted of an electroluminescent foil, a square-shaped rotary contact and a DC motor. The motor was fixed to the acrylic structure of the system. Through a metal axis, the rotary contact and the electroluminescent foil were attached to the motor. Two fixed open wires powered the foil through the rotary contact. The visual impact was affected by both the revolving speed of the motors and the on/off state of the electroluminescent strips.

Insomnia (Fig. 4.98) focused on the flexibility, thinness and consistent illumination of the material. These properties were used to backlight two optical animations based on moiré patterns. The structure consisted of an electroluminescent layer, printed black and white pictures, and a striped pattern, which slid horizontally. The lightness of the first object, a hovering cylinder with a total thickness of only 6 mm, resulted from the gauge of the electroluminescent foil (0.3 mm) and its extreme transformability. The electroluminescent sheet was attached to the inside of a 5-mm MDF frame and covered by opaque and transparent layers. The structure could be entered from below and formed a 360° animated surface. The second object used a 9-m long continuous red electroluminescent tape of 2 cm width, which sat below the suspended cylinder. The form resembled a cloverleaf and housed a small letter-based animation in each corner.

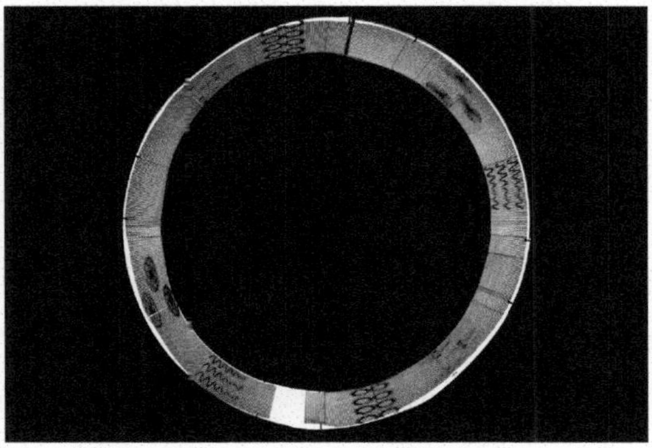

Fig. 4.98 Insomnia was a large-scale optical animation that slowly rotated around a person standing at its center (Kretzer, Material Animation 2011)

Each room was equipped with a Microsoft Kinect Camera, which sensed the amount of people, their location, movement and speed, and if more than two people were in a room, also the area they occupied (Fig. 4.99). Every installation was linked to a computer that communicated with a remote server, which was responsible for data distribution and overall performance of the spaces. After the information was compiled individual data packages were sent back for interpretation on the local PC. The behavior of the rooms and the installations they contained was organized in two layers. The first layer was dependent on the current amount of people in the room, whereas the next layer was related to their average speed. Different behaviors were designed to encourage visitors to further engage with the installation while getting a real-time feedback on their actions within the space. Serial communication was used to send signals from the remote network to decentralized Arduino Microcontrollers, which in turn controlled servo and DC motors to move the elements and transistor/relay circuits to switch the electroluminescent foils.

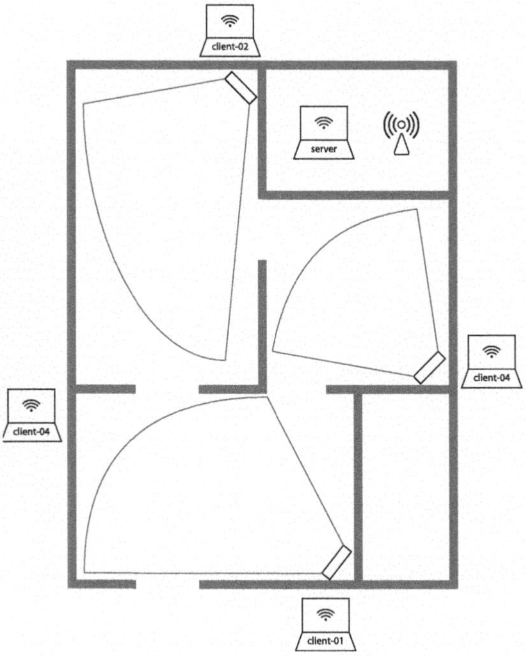

Fig. 4.99 Floor plan of three interconnected rooms and the location of cameras and computers (Christensen, Diagram of Material Animation Layout 2011)

4.6.4.2 Actuated Matter (2011)

The *Actuated Matter* workshop took place at the Zurich University of the Arts in July 2011 and was co-organized by Karmen Franinovic, Daniel Bisig, Mathias Gmachl and Rachel Wingfield. In addition to the organizers, the event included 20 participants from a variety of backgrounds and disciplines. The core aim was to realize a prototypical environment, which would dynamically respond to the presence and behavior of its inhabitants. The immediate reactions were of kinetic as well as visual, acoustic, and sensual nature. This was achieved through the combination of different smart materials with lightweight construction methods and flexible skins and the integration of behavioral processes of self-organization to augment the system with properties of autonomous organisms.

During the 5 days workshop the participants developed a speculative model for membrane structures that exhibited properties of sensitivity, resilience, and decay. The project followed a hands-on approach and encouraged the participants to physically engage with the intrinsic properties of various smart materials (Fig. 4.100). The active elements included electroluminescent screens, electroactive polymers, and thin-film audio panels. The electroluminescent displays were printed onto flexible ITO-coated substrates. The structural system was based on a three-dimensional network of interlinked loops, made from optical fibers. This structure, which reminded of early experiments by Yona Friedman, formed a lightweight system with adequate flexibility to become activated but enough rigidity to support a number of responsive elements (Fig. 4.101).

Fig. 4.100 Installation in the making (Kretzer, Actuated Matter 2011)

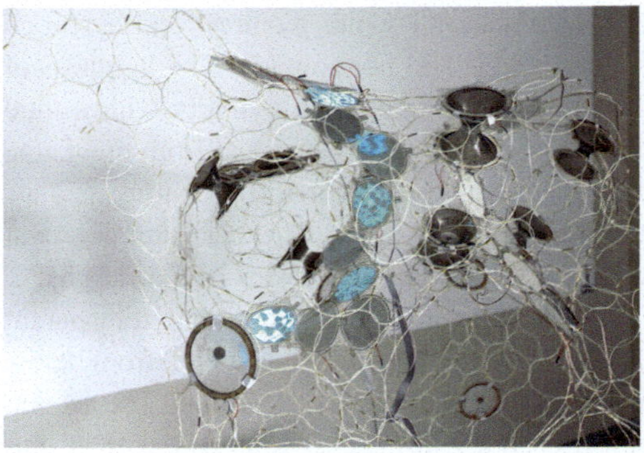

Fig. 4.101 Final structure containing luminous electroluminescent displays, DIY magnet film speakers, and moving electroactive elements (Kretzer, Actuated Matter 2011)

Driven by their own curiosity and engagement the participants, who had little or no previous knowledge in working with such materials, had a lot of freedom in designing and creating customized and individual components. The workshop proved that highly sophisticated materials and structures could emerge from participatory and collaborative strategies (Franinovic et al. 2013, p. 213).

4.6.4.3 Luminous Textiles (2015)

Luminous Textiles (Fig. 4.102) was a workshop held at the Swedish School of Textiles, University of Borås in September 2015. During, the 1 week project a range of flexible electroluminescent displays and luminous surfaces was produced. In groups the participants explored techniques in screen-printing and controlling the devices digitally. In addition to experimenting with various shapes and patterns the most interesting aspects involved printing with differently colored phosphors. Methods in printing straight onto textile substrates were also researched, however due to insufficient conductivity of the base material and problems in isolating the electrodes properly remained largely unsuccessful. Another difficulty remained cleaning the screens from the oil-based inks. Layering various colors on top of each and mixing them into 'new' colors was truly inspiring and worth investigating further.

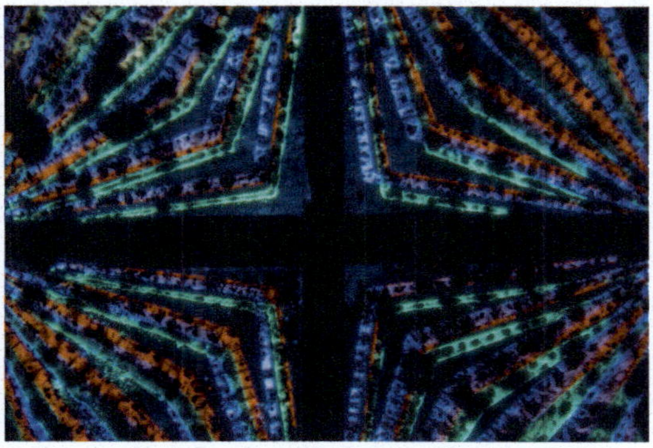

Fig. 4.102 Printing various colors on top of each other was one of the key developments achieved during the workshop (Kretzer, Luminous Textiles 2015)

4.6.5 Applications

Electroluminescent displays are today used in a wide variety of applications. They are made to illuminate liquid crystal displays (LCD), advertising boards, automotive panel backlights, or safety-related products. As consumer materials they can be obtained in the form of electroluminescent strips, cables, wires, inks, or films, and sheets.

In an architectural application electroluminescent displays first emerged in 1957 as luminous surfaces in a model room. The 112 square panels were based on a sandwich structure of rigid sheet metal plates that were covered with a thin layer of porcelain to which the electroluminescent film was attached. The panels covered the halves of three walls and a ceiling (Ritter 2007, p. 132). Around the same time small flat electroluminescent panels became popular as night-lights or optical animations, which could be plugged into any available power outlet (Science Service Smithsonian 1961).

Today electroluminescent displays are used both in interior and exterior applications. But even though they are being industrially produced in large numbers and continuously increasing size, surfaces at an architectural scale can still only be achieved when grouping several smaller electroluminescent displays together.

Electroluminescent displays are being applied as luminescent floor or wall coverings or large format displays in public spaces. Various artists have experimented with EL surfaces, including Dutch designer Jonas Samson, who developed electroluminescent wallpaper that can display simple animations and patterns. A similar concept is now commercialized by the electroluminescent store and supposedly allows users to create their own illuminated decorating walls that can be controlled with mobile phones or tablets (The Electroluminescent Store).

London-based Loop.pH design studio developed in 2004 their ornamental 'Blumen' electronic wallpaper display in the format of sliding panels. The repetitive pattern allowed the piece to be cut in smaller parts that could then be rejoined into different assemblies. The floral patterns were printed on a textile base and responded to changes in the environment based on the particular space and the integrated sensors of choice (Loop.ph).

In 2007, Spanish architect Santiago Calatrava collaborated with the Italian furniture maker Poltrona Frau to develop an illuminated seating system for the four large theaters and the small rehearsal space of his Palau de les Arts Reina Sofia, the opera house in Valencia. The company produced and installed 4219 seats with integrated electroluminescent films to be used as background illumination and emergency guidance. Depending on the preferred atmosphere the chairs could then be variably powered from 50 to 150 V, creating a feeling of weightless floating in the dark (Poltrona Frau).

4.7 Electroactive Polymers

Electroactive polymers (EAP) are polymers that change their size, shape, or volume in response to a strong electrical field, which creates a force on charged particles (Fig. 4.103). In the domain of shape-changing materials (e.g., piezoelectrics, thermo-elastic polymers, shape-memory alloys and polymers, or magnetostrictive materials) electroactive polymers stand out due to their large active deformation potential, high response speed, low density, and improved resilience. They are extremely lightweight, inexpensive, and compliant and can also be used as stretch sensors.

According to their working principle they can be split into two main categories: ionic and electronic electroactive polymers.

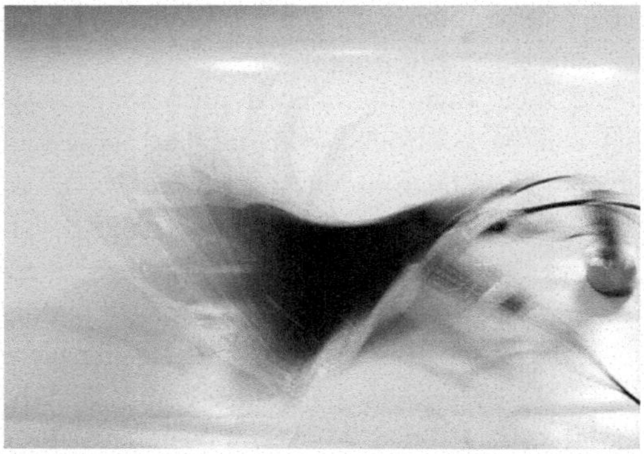

Fig. 4.103 Electroactive polymer during activation (Kretzer, Electroactive Polymer 2012)

- Ionic electroactive polymers are driven by a displacement of ions during electrical stimulation, which leads to a change in shape or volume. Their main advantage is that they can be actuated by voltages as low as 1–2 V, however, since the ions are diffused inside an electrolyte they need to maintain their wetness during all times. They are mostly used as bending actuators with strong bending capabilities where high forces are required but have a rather slow response speed. Since the production of stable material configurations requires high precision, they are expensive and usually not commercially available.
- Electronic electroactive polymers are driven by strong electric fields. The occurring electrostatic forces lead to an electromechanical change in shape of the material. Usually, they are applied as planar actuators due to their large in-plane deformations. In contrast to ionic electroactive polymers, they work in dry conditions but need very high activation voltages in the range of several kilovolts. They exhibit a very short response delay, display a relatively large activation stress, and can hold the induced displacement under DC activation (Bar-Cohen 2002, p. 2).

4.7.1 History

The first reported occurrence of an electroactive phenomenon dates back to 1880, when Wilhelm Röntgen observed a length-change in a fixed rubber band with a weight on its one end that was electrically charged and discharged. In 1899, M.P. Sacerdote confirmed Röntgen's experiment and formulated a theory of the strain response to electric field activation. The next milestone was in 1925, when the first piezoelectric polymer, called Electret, was discovered by combining carnauba, wax, rosin, and beeswax, and then subsequently cooling and solidifying the mixture while exposing it to a DC bias field (Bar-Cohen 2004, p. 9).

In 1969, it was demonstrated that polyvinylidene fluoride (PVDF) displays a large piezoelectric constant and this discovery hence launched research into polymers that would exhibit a similar effect. The first electrically conducting polymers were discovered in 1977 by Hideaki Shirakawa by doping polyacetylene with iodine and thus enhancing its conductivity by eight times, close to that of metal (Finkenstadt 2005, p. 735). In the early 1990s, ionic polymer-metal composites were developed, which displayed a much lower activation voltage while sustaining a larger deformation than previous electroactive polymers.

In order to promote the potential of electroactive polymers an arm-wrestling match between an EAP powered robotic arm and a human was posed at the opening of the first *Electroactive Polymer Actuators and Devices (EAPAD) Conference* in 1999. Three international research institutes accepted the challenge, including Environmental Robots Incorporated (ERI), the Swiss Federal Laboratories for Materials Testing and Research (EMPA), and Virginia Tech. The first match was held at the EAPAD Conference in San Diego, California in 2005, and while the human opponent easily won against all competitors, the device from ERI was, with 26 s, the longest to last (Bar-Cohen 2005, p. 3).

While EMPA research had focused on spring-roll actuators for their arm-wrestling robot, in 2007 Patrick Lochmatter presented the development of a shell-like electroactive polymer actuator, which was capable to perform complex out-of-plane deformations (Lochmatter 2007, p. 23). The research was taken further and used to propel an 8 m long and 2 m high airship by adding large-scale electroactive polymer devices to its sides and rear fin (Jordi 2011, p. 113).

The first commercially developed device containing electroactive polymers as artificial muscles was a fish, being able to swim on its own, produced by Eamex in Japan in 2002. It worked without batteries or a motor, simply relying on electroactive polymer materials that bend when energized through inductive coils at the top and bottom of the fish tank (Bar Cohen 2004, p. 4). In 1990, DARPA funded research at SRI International, which led to the development of an electroactive polymer based on silicone and acrylic polymers. The company Artificial Muscle further developed the technology in 2005 and started industrial production in 2008. In 2010, Artificial Muscle became a subcompany of Bayer Material Science (SRI International). Since 2008, Danfoss manufactures and commercializes a dielectric electroactive polymer material called Polypower, which is mostly intended for use as variable stretch sensors.

4.7.2 Material Structure and Operation

Among electronic electroactive polymers especially soft dielectric EAP also referred to as dielectric elastomers (DE), show promising functional properties since they are capable of strains up to 380 %, extremely flexible, light, thin, transparent, and can basically be tailored to any size or shape (Ritter 2007, pp. 66–69).

Figure 4.104 shows the working principle of a dielectric elastomer actuator. The central element of dielectric elastomer actuators consists of a thin elastomeric film (2), which can be silicone or acrylic. The membrane is coated on both sides with or sandwiched between two compliant electrodes (1, 3). In this configuration, the polymer acts as a dielectric in a compliant capacitor (Jordi 2011, p. 15).

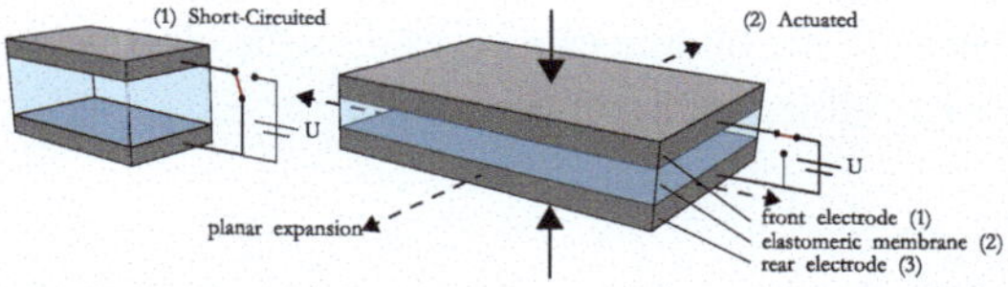

Fig. 4.104 Basic dielectric elastomer working principle: when voltage is applied the elastomeric film gets compressed in thickness leading to a planar expansion in area (adapted with permission from Michel, Empa Material Science and Technology, Basic dielectric elastomer working principle 2010)

When an electrical voltage is applied between the electrodes, the opposite charges move from one electrode to the other, and hence squeeze the film in its thickness direction. Since the material is close to incompressible this leads to a planar film expansion. When the voltage is turned off and the electrodes are short-circuited the electroactive polymer returns to its original shape. The observed response of the polymer film is primarily a result of the interaction between the electrostatic charges on the electrodes. Based on the principle of operation of soft dielectric elastomer, two main directions to perform work against external forces are possible and as such two main types have been developed (Lochmatter 2007, pp. 12–13).

- The expanding actuators perform work against external pressure loads in (both) planar directions. They are usually under a certain pre-stretch and are often embraced by a

mechanical support structure or frame. In these systems, the support is released, when the material relaxes under activation. Depending on the pre-stretch ratio different expansions in varying directions can be achieved.
- The stacked or contractile actuators consist of several hundred or thousand layers and can work against external tensile loads acting in thickness direction, since the material is compressed when activated. They are produced from non-stretched material and can transmit the force directly onto an object (Jordi 2011, p. 23).

4.7.3 Fabrication

In order to successfully create a dielectric elastomer membrane, which is capable of out-of-plane deformations as shown in Figs. 4.105 and 4.106 both the membrane film as well as the electrodes must fulfill certain properties.

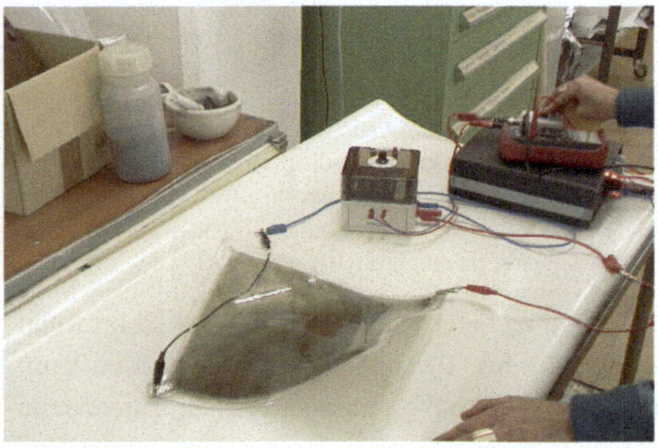

Fig. 4.105 Electroactive polymer actuator in relaxed state (Kretzer, Electroactive Polymer 2010)

Fig. 4.106 Electroactive polymer actuator in active state (Kretzer, Electroactive Polymer 2010)

The film must be thin with uniform thickness, incompressible and a good electrical insulator, i.e., have high electric breakdown strength. Silicones, polyurethanes, and acrylic-based polymers have emerged as most suitable materials. Silicones show fast activations responses and can be operated in a large temperature range (Lochmatter 2007, p. 16). Polyurethanes exhibit larger force output than silicones and acrylics feature a good dielectric strength (Jordi 2011, p. 20). Among the commercially available acrylics in particular VHB (very high bond) films by 3M perform great in terms of electromechanical strains, producing high pressures and extraordinarily high specific elastic energy densities. When the stretched film (3) is attached to a flat, flexible frame (6), a minimum energy structure results, when the actuator is in its relaxed state. The complex shape (Fig. 4.107) arises from the interaction between the different stress and strain fields between membrane and frame (Kofod et al. 2006, p. 141).

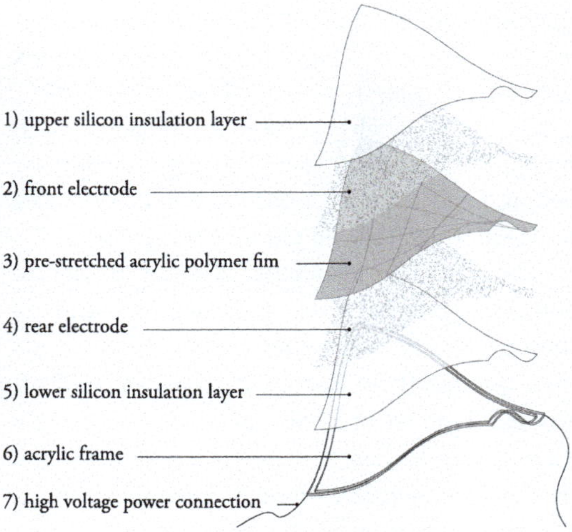

Fig. 4.107 A basic dielectric elastomer consists of three layers, a front electrode, a compliant membrane, held by a flexible frame and a rear electrode. For reasons of safety or durability both front and back can be insulated (Kretzer, Electroactive Polymer 2010)

The electrodes (2, 4) for the application on dielectric elastomer transducers need most of all feature good conductivity, even when stretched. They must be as thin as possible and bond well with the dielectric film. This is usually achieved by the application of electrically conductive particles, such as metal, carbon black, or graphite powder. The particles can be mixed with grease, oil, or liquid silicone for thicker electrodes with lower electrical resistance and are applied by spraying or brushing. An extra layer of silicone can be used to insulate the dielectric elastomer component and increase its longevity and safety (1, 5). In order to supply, the dielectric elastomer with the necessary DC activation voltage high voltage sources or amplifiers are utilized (7). In cases where activation time is not crucial, boost converters can be used instead, which are less bulky and expensive.

4.7.3.1 Materials and Tools

See Fig. 4.108.

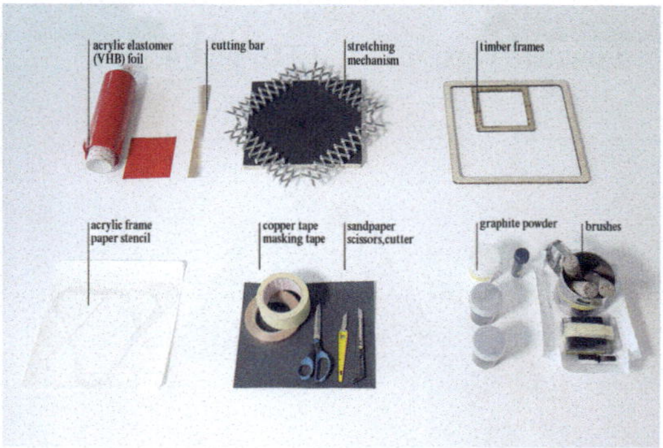

Fig. 4.108 Elastomeric membrane foil (VHB 4910 or VHB 4905)/cutting bar, ruler, stretcher/timber frames to maintain pre-stretch/laser-cut acrylic frame/paper stencil, sandpaper, scissors, cutters/aluminum foil/conductive copper tape/scotch tape/graphite powder carbon black/rubberpads, brushes/5–5000 V power converter (G 50R or G 60R)/high voltage resistors (4 × 10 MΩ)/ laser cutter (Roman and Pantazis, Electroactive Polymer Tutorial 2012)

4.7.3.2 Steps

Shape design

The desired shape can be designed using a graphical computer program that exports laser-ready files. Rhombic shapes have proven to work nicely, but other designs are possible, too. The frame is then cut from acrylic, polystyrene, or polycarbonate using a laser cutter. Other plastics work as well but the flexibility/stiffness ratio has to comply with the force of the dielectric elastomer. The paper stencil can also be cut with the laser cutter. It should have an internal offset of about 5 mm smaller than the plastic frame. A number of variables and their interplay determine whether the design will work and actually move:

- the size of the frame shape
- the thickness of the plastic material (usually between 0.5 and 1 mm)
- the width of the frame border
- the amount of pre-stretch (the higher the pre-stretch the more the component will move)
- the type of frame material.

The whole process is still very much based on trial and error and empirical shape development.

Preparation (Fig. 4.109).

Fig. 4.109 A clean and well-prepared surface is important since any type of dust or dirt on the membrane could lead to electrical breakdown of the components. All tools and materials should be at hand (Roman and Pantazis, Electroactive Polymer Tutorial 2012)

Membrane (Figs. 4.110, 4.111 and 4.112).

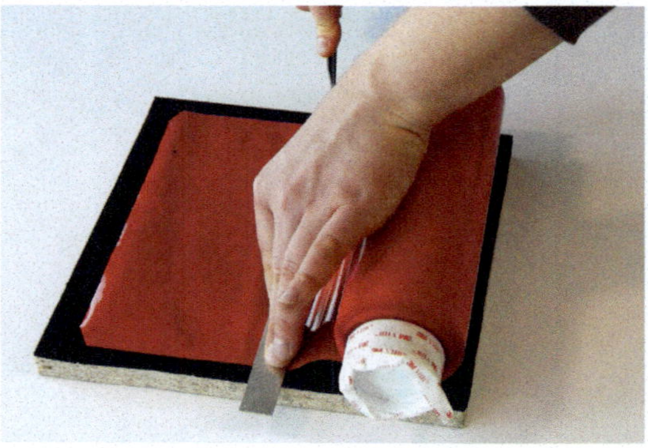

Fig. 4.110 The membrane needs to be cut from a larger roll of VHB and should have roughly the same size as the stretching mechanism in its folded state (Roman and Pantazis, Electroactive Polymer Tutorial 2012)

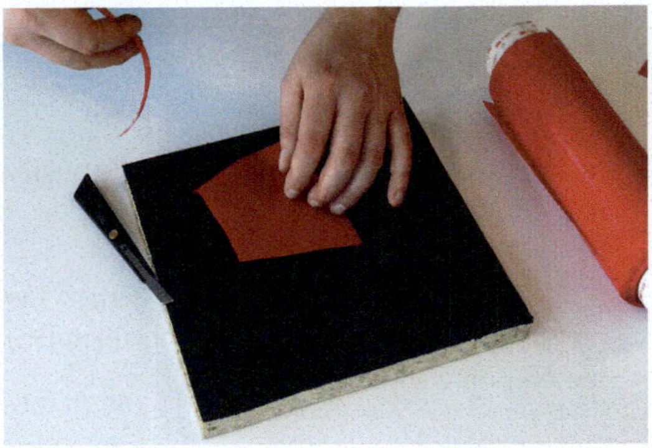

Fig. 4.111 Small strips of the material should be added to the borders of the membrane for reinforcement (Roman and Pantazis, Electroactive Polymer Tutorial 2012)

Fig. 4.112 The membrane is then fixed onto the stretching mechanism. All borders need to be attached thoroughly to the stretching mechanism to avoid loosening during stretching (Roman and Pantazis, Electroactive Polymer Tutorial 2012)

Stretching (Figs. 4.113 and 4.114).

Fig. 4.113 After removing the protective layer of the film, the membrane is slowly but consistently expanded in all directions. The stretch-ratio will affect the later movement. It depends on the thickness of the membrane but is usually around three to five times of the original size. This is a delicate process and bit of training might be necessary (Roman and Pantazis, Electroactive Polymer Tutorial 2012)

Fig. 4.114 Finally the membrane is fixed with masking tape onto the static wooden frame, which will maintain the pre-stretch (Roman and Pantazis, Electroactive Polymer Tutorial 2012)

Powder (Figs. 4.115, 4.116 and 4.117).

Fig. 4.115 After placing the stencil foil onto the membrane the conductive powder can be applied (Roman and Pantazis, Electroactive Polymer Tutorial 2012)

Fig. 4.116 The carbon powder needs to be rubbed firmly but smoothly into the material (Roman and Pantazis, Electroactive Polymer Tutorial 2012)

Fig. 4.117 Once one side is done the stencil can be removed and the procedure repeated on the backside. Small lines for the electronic connectors should be printed onto each side on positions not opposing each other (Roman and Pantazis, Electroactive Polymer Tutorial 2012)

Electronic connection (4.118).

Fig. 4.118 A small piece of conductive tape or aluminum foil should be fixed on *top* of the previously marked lines on both sides of the membrane (Roman and Pantazis, Electroactive Polymer Tutorial 2012)

Plastic frame (Figs. 4.119 and 4.120).

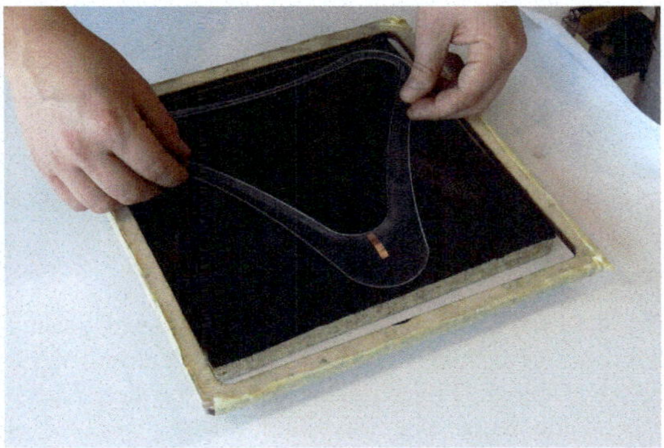

Fig. 4.119 The plastic frame needs to be firmly pressed onto the membrane. Once all sides are attached equally it can be cut using a knife or scalpel (Roman and Pantazis, Electroactive Polymer Tutorial 2012)

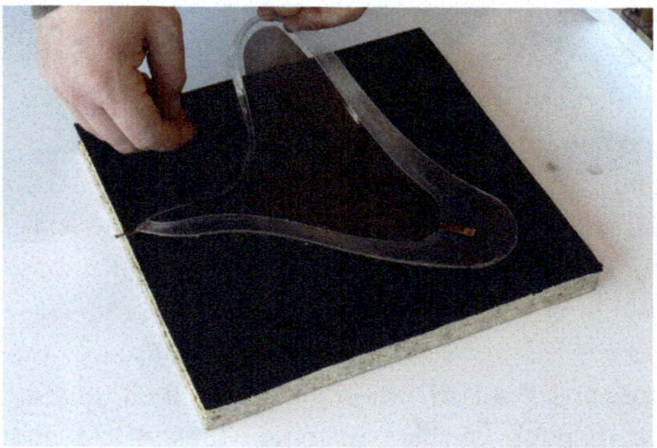

Fig. 4.120 The frame will then relax and take its deformed state (Roman and Pantazis, Electroactive Polymer Tutorial 2012)

Power (Fig. 4.121).

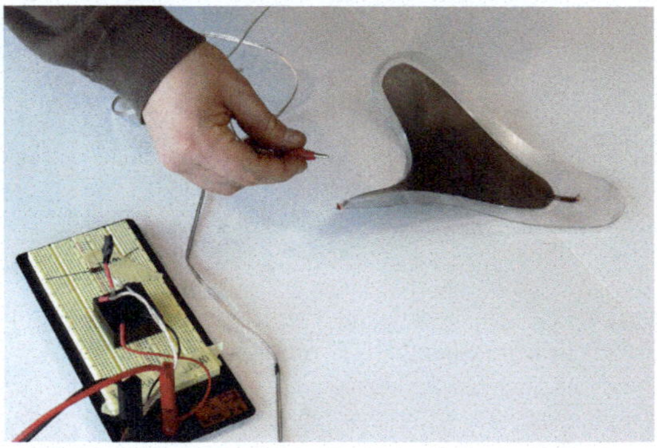

Fig. 4.121 After the electrodes have been connected to a 5000 V (5 kV) power converter the element will slowly move into its activated state. A potentiometer attached to the power source will help to more precisely control the output voltage (Roman and Pantazis, Electroactive Polymer Tutorial 2012)

4.7.4 Case Studies

Since the aesthetic properties and the soft organic movement of dielectric actuators are highly intriguing, the material has been used in a number of projects that were realized as part of this research. The sole focus on dielectric elastomer actuators has however only been during *ShapeShift*, a final group thesis at the Chair for CAAD in 2010 and *Animated Textiles*, a workshop at the Swedish School of Textiles, Boras in late 2012.

4.7.4.1 ShapeShift (2010)

ShapeShift was the final MAS CAAD thesis project of Edyta Augustynowicz, Sofia Georgakopoulou, Dino Rossi, and Stefanie Sixt and was realized in September 2010. The project investigated the use of dielectric elastomers to create dynamic spatial applications. While the main application of dielectric elastomers is to produce artificial muscles, the students focused on highlighting their quality as dynamic surface material (Figs. 4.122 and 4.123).

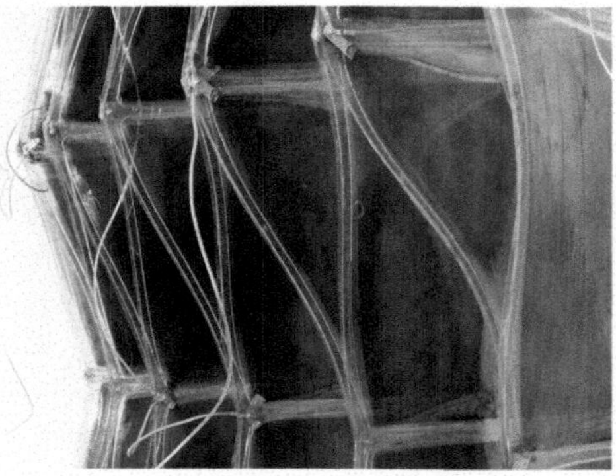

Fig. 4.122 The installation ShapeShift was exhibited in Zurich in September 2010 (Kretzer, ShapeShift 2010)

Each of the 36 elements within the structure consisted of a pre-stretched film that was attached to flexible acrylic frames and sandwiched between to compliant electrodes. Once a high DC voltage (3–5 kV) was applied the film was compressed in its thickness direction, leading to a planar expansion. Since the membrane was attached to the flexible acrylic frame, the frame bent when the material was in its relaxed state and flattened out when the tension was removed during actuation (Kretzer 2012, p. 81).

Fig. 4.123 The final system consisted of 36 elements, which were actuated, in unison (Kretzer, ShapeShift 2010)

Through continuous design iterations the students altered the acrylic frames until the movement was maximized and the desired three-dimensional motion was achieved. In parallel to the development of single components investigations into structural arrangements were performed. Through connecting a multitude of components together dynamic configurations could be achieved that enhanced the movement even further and resulted in feasible self-supporting structures (Fig. 4.124). Similar to the single shapes the final form of these tessellations resulted from the relationship of the dielectric elastomer to its frame and the connections to neighboring elements (Kretzer and Rossi 2012, pp. 480–481).

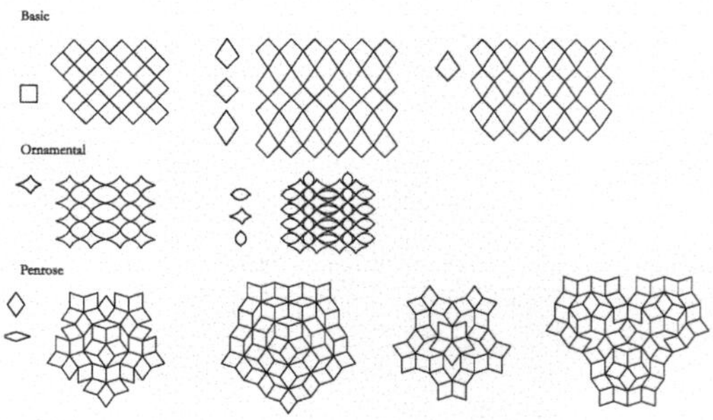

Fig. 4.124 Various tiling options during development. From *top* to *bottom*: basic, ornamental, Penrose (MAS CAAD Diploma Project, Edyta Augustynowicz, Georgakopoulou, Rossi, Sixt, ShapeShift 2010)

4.7.4.2 Animated Textiles (2012)

The use of dielectric elastomers was further investigated during a one-week workshop in fall 2012, held together with Ivana Damjanovic and invited by Delia Dumitrescu at the Swedish School of Textiles in Boras. During this workshop the thirteen participants explored in groups of four the combination of dielectric elastomers with various lightweight textiles in order to create animated surfaces, structures, and assemblies (Fig. 4.125).

After a basic two-day introduction into the art of producing the material the students started experimenting with different shapes, based on previously prepared, working designs. They then developed their designs, which were produced from laser-cut polycarbonate sheets. The designs, which exhibited the most promising deformation and actuation, were further tuned and combined with a variety of fabrics, knitted structures, or textiles.

Fig. 4.125 Combining custom electroactive polymers with lightweight fabrics was one goal of the Animated Textiles Workshop (Kretzer, Animated Textiles 2012)

The empirical study of how much force the membranes were able to produce, how this motion could be enhanced through textile elements, and how these could then become visually appealing systems marked the core tasks of the workshop (Fig. 4.126). At the end each group had built a prototype, which was shown to a larger audience. The participants who were mostly design and textile students both on Master and PhD level were intrigued by the soft and organic movement of the components and speculated how it could be used as responsive garments or interactive clothing (Kretzer 2013a, b, p. 618).

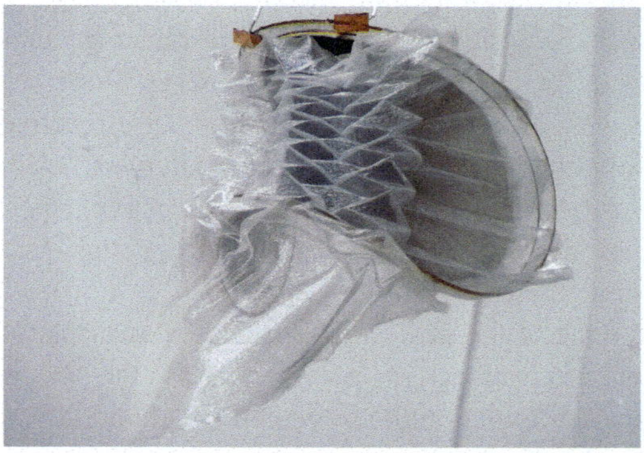

Fig. 4.126 Pleated textiles, attached to the electroactive membranes, lead to interesting shapes and intricate movements (Kretzer, Animated Textiles 2012)

4.7.5 Applications

Despite decades of research and continuous progress the domain of electroactive polymers is still far from mature and a number of topics, such as long-term performance and durability need much further development. Nevertheless, a number of potential applications have emerged mostly based on the shape flexibility of electroactive polymers, which among other qualities, makes them very versatile materials. They have been utilized as artificial muscles, both for prosthetic purposes or 'super-human' exoskeletons, microelectromechanical systems (MEMS), medical devices, speakers, and more recently, as refreshable braille displays (Runyana and Blazieb 2010, p. 764207-3), or gaming interfaces (Evangelho 2014). Based on their immediacy and 'natural' or 'organic' behavior recent research has investigated their potential for active camouflaging on soft surfaces, mimicking cephalopod chromatophores (Rossiter et al. 2012, p. 5). But even though a large variety of electroactive polymer actuators have been successfully demonstrated most materials are still custom made and the number of commercially available products is limited.

Besides using them as mechanical actuators they can also perform as sensors since when the charged device is stretched or contracted, the serial resistance within the electrodes changes. Furthermore, they can be used to convert mechanical work into electrical energy and as such be applied as dielectric elastomer generators, for example to harvest energy from sea waves. Other types of electronic electroactive polymers include ferroelectric polymers, electrostrictive graft polymers, electrostrictive paper, piezoelectric polymers, and liquid crystal elastomers (Bar-Cohen 2002, p. 2).

One of the earliest design-related concepts of using electroactive polymers was the thesis project 'Funktionide' by Stefan Ulrich in 2009, a speculative future object that wants to satisfy emotional needs, practically exemplified by a number of physical experiments with electroactive polymer actuators (Ulrich 2009, p. 48). More likely however will be the use of electroactive polymers for portable devices and peripherals, with a predicted penetration of the haptics for consumer electronics touch display market exceeding 60 % until 2018, which will account for more than 40 % of the expected total revenue in 5 years (IDTechEx).

For architectural applications electroactive polymers exhibit very promising properties, especially their homogeneous surface quality, transparency, and large active deformation are visually astonishing. Problematic is however their scalability, since fabrication is still largely manual and potential errors dramatically increase once reaching a certain size. Furthermore, the durability of the material cannot yet comply with architectural standards especially when exposed to changing environmental and climatic conditions. In open spaces also the accessibility and location of the material needs to be considered, since the components need fairly high voltages, which pose a certain danger. Thus, it would be a requirement to insulate publicly accessible elements

The New York-based architectural office DeckerYeadon proposed in 2011 to apply electroactive polymers to a double-glazed homeostatic facade system to regulate heat loss and heat gain by opening and closing autonomously in response to changes in interior room temperature (DeckerYeadon). A similar concept was developed by Hickok Cole Architects from Washington in 2013 for their vision of the 'Office Building of the Future' at *NAIOP's Development 2012 Conference*, which employed a dynamic electroactive polymer shading layer activated by the sun (Architizer).

Aurélie Mossé at the Center for Information Technology and Architecture (CITA), Copenhagen investigated in her project Reef how adaptive minimum energy structures in the interior can be applied to re-establishing a reciprocal relationship with nature. The electroactive modules, which populated a ceiling, opened and closed in response to wind intensity and direction (Mossé et al. 2011, p. 9).

4.8 Soft Robotics

Soft Robotics is a research area that draws inspiration from **invertebrates** (animals that do not have internal skeletons) to develop alternatives to hard bodied robots, which consist mainly of rigid and heavy metal parts. While conventional systems have a long tradition and various advantages, like reliability, precision, strength, and durability, they are also facing many limitations that seem hard to overcome with predominant technologies. Some of these restraints are based on their mechanical complexity, like the agitation in difficult terrain or the handling of fragile objects, whilst others come from their weight, limited ranges of motion of standard actuators, or the costs involved in producing such systems. In soft robotics the potential of compliant soft materials that allow for continuous deformation are explored to create structures whose ranges of motion are directly defined by the properties of the materials used in their fabrication. Through this shift the field of robotics becomes more closely related to materials science and chemistry than to mechanical engineering.

4.8.1 Material Structure and Operation

While this interest bridges various areas, like the development of electroactive polymers (EAP), shape-memory polymers (SMP), or pneumatically driven McKibben-type actuators, generally referred to as air muscles, one particular focus is on the creation of pneumatic networks embedded in silicon-based elastomers of variable stiffness (Fig. 4.127).

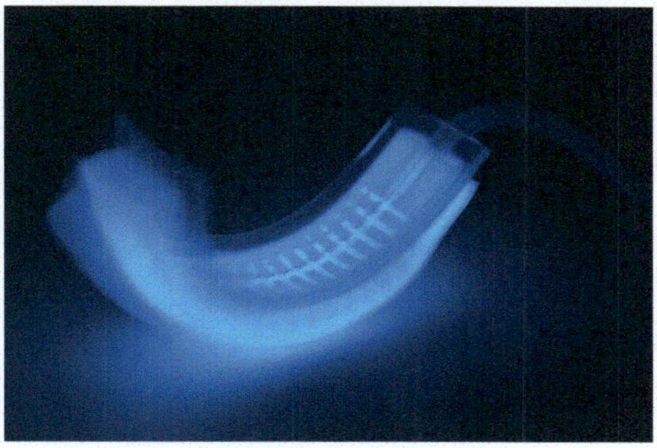

Fig. 4.127 Phosphorescent soft robotic actuator during activation (Kretzer, Soft Robot 2012)

The structure of such a system is fairly basic (Fig. 4.128). It usually consists of a soft elastomeric layer (1), which can be produced from soft silicone in a cast molding technique. Embedded into the layer is a network of air channels (3), potentially separated into various chambers, which are inflated through a series of air inlets (2). The amount of air inlets depends on the number of independently actuated areas. The soft layer and the air inlets, which can be silicone tubing, are attached to a strain-limiting layer (4) made from a harder silicone.

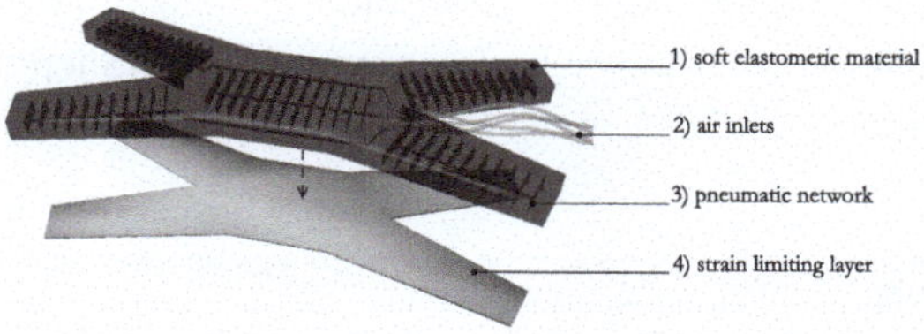

Fig. 4.128 Schematic representation of basic soft robot structure (Shepherd et al. 2011, Fig. 1)

The technology, which was originally developed at the Department of Chemistry and Chemical Biology and the Wyss Institute for Biologically Inspired Engineering at Harvard University, draws continuous inspiration from nature and has investigated a number of different potentials.

4.8.1.1 Locomotion

Soft organisms, many of which are found in marine environments, are highly versatile locomotors since they do not have the need for endo or exoskeletons like land animals that have to retain form in a gravitational field. Trying to mimic some of the motions and features of such organisms, soft elastomeric robots use embedded pneumatic networks and chambers that inflate during actuation. The combination of layers of differently extensible materials and the resolution, size, and arrangement of the internal network can then be used to create bending motion. The researchers at Harvard University have thus constructed a tetrapod that can lift each of its four legs independently. The embedded pneumatic networks are pressurized through computer controlled external sources, which leads to crawling and undulating movement (Fig. 4.129). The nonlinearity in the motion of the single elements can produce very complex actuation by relying on a rather basic control system (Shepherd et al. 2011, p. 20401).

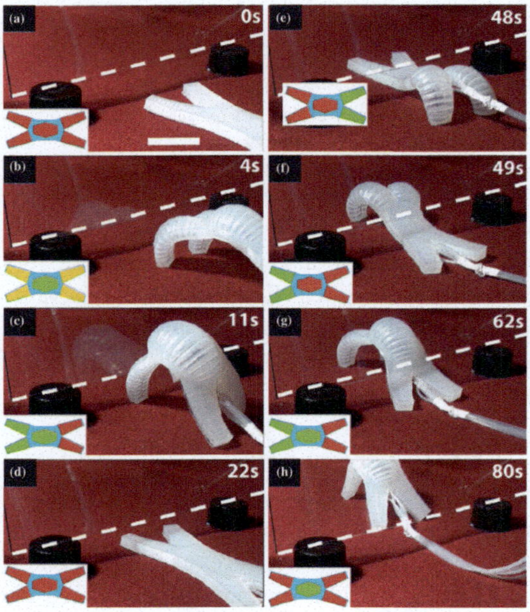

Fig. 4.129 Snapshot of a moving soft robot (Shepherd et al. 2011, Fig. 4)

4.8.1.2 Gripping

To further prove the potential of active soft structures, researchers from the same group have produced a starfish-like, six legged gripper, consisting of an active compliant layer, a closing layer made from a stiffer silicone, and a third textured layer to enhance the gripping effect. The actuated legs of the gripper curl downward and are able to carry various objects without damaging or harming them, since the soft interface distributes the force over the complete contact area (Fig. 4.130). The experiment shows that soft robots have high potential for applications that require handling delicate objects and where the implementation of feedback sensors would be too difficult or expensive (Ilievski et al. 2011, p. 1894).

Fig. 4.130 Soft-robotic gripper grabs chicken egg (*above*) and mouse (*below*) (with kind permission from John Wiley and Sons from Ilievski et al. 2011, p. 1893, Fig. 5)

4.8.1.3 Camouflage

Again drawing inspiration from animals, like squid, cuttlefish, chameleons, or fireflies that can dynamically change their body patterns and colors for reasons of disguise, protection, deterrence, display, or communication, a network similar to the one used for inflation, can allow the transportation and emergence of colored fluids within the soft machine. Even though this system is obviously much less complex than the processes happening in color-changing organisms, it allows for relatively fast changes, a large variety of color, the cover of large areas, and the transport of both liquids and fluids. Furthermore, since the color layers are very light they do not constrain the movement of the robot (Morin et al. 2012, p. 829).

4.8.1.4 Immediacy

Even though pneumatic expansion of a network of channels embedded in soft elastomers provides a simple method to create complex movements, the speed of actuation is relatively slow, due to the time the air needs to fill and inflate the micro channels. A possible solution is the creation of explosive bursts within the network, using a chemical reaction. In particular, the explosive combustion of hydrocarbons, triggered by an electrical impulse allows a tripedal robot to jump 30 times its body height in less than 0.2 s. By improving the timing of actuation, the jump height can be further increased, the energy efficiency enhanced, and the direction controlled (Shepherd et al. 2013, p. 2966).

4.8.2 Fabrication

Since soft robotics is a fairly new area of robotics, methods for design and fabrication are much less developed and accessible than for hard robots. Materials that are used to fabricate soft robotic devices are usually polymers (especially elastomers), which fall into the discipline of organic materials science. The fabrication process is relatively simple and quick. The phosphorous powder that is used in the following tutorial is additional and does not influence the working principle of the device; it merely makes it visible in the dark.

4.8.2.1 Materials and Tools

See Fig. 4.131.

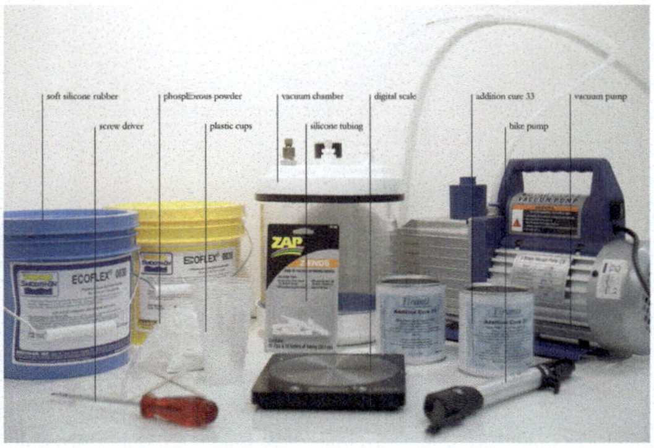

Fig. 4.131 Soft silicone rubber (Eco Flex 00-30)/vacuum chamber/plastic cups/silicone tubing, tube z-ends/hard silicone rubber (Addition Cure 33)/phosphorous powder/screwdriver, scale, bicycle pump/gloves, scissors, cutter, kitchen paper/3D printed mold (Kretzer, Soft Robots Tutorial 2012)

4.8.2.2 Steps

Mold making (Fig. 4.132).

Fig. 4.132 The mold for the soft robot can be created using 3D CAD software. It needs to include the desired capillary system and can then be printed using an ABS or similar 3D printer (Kretzer, Soft Robots Tutorial 2012)

Soft silicone mixture (Figs. 4.133, 4.134 and 4.135).

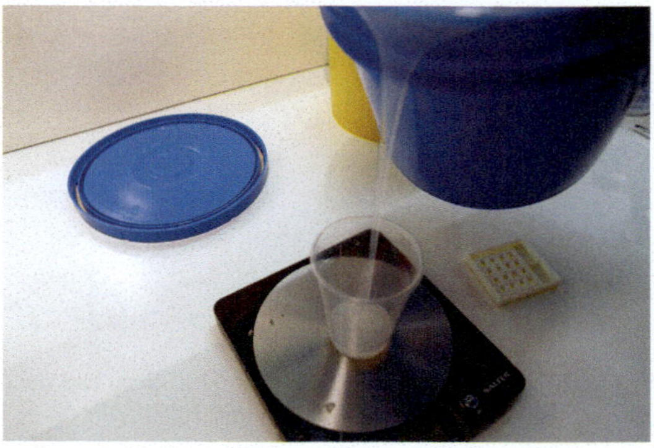

Fig. 4.133 The Ecoflex Supersoft silicone rubber part A and part B have to be mixed in a 1:1 ratio according to the volume of the mold (Kretzer, Soft Robots Tutorial 2012)

Fig. 4.134 Continuous stirring until both parts are mixed equally is important (Kretzer, Soft Robots Tutorial 2012)

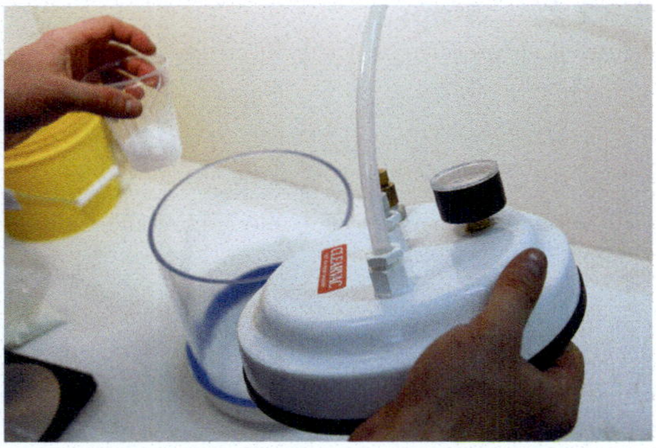

Fig. 4.135 Next the solution is degassed in the vacuum chamber until no more bubbles appear. The phosphorous powder can then be added in a ratio of 1 (powder) to 5 (silicone). The solution again needs to be stirred and degassed (Kretzer, Soft Robots Tutorial 2012)

Soft part of robot (Figs. 4.136 and 4.137).

Fig. 4.136 The mixture is poured into the clean and dust free mold (Kretzer, Soft Robots Tutorial 2012)

Fig. 4.137 Once the silicone has evenly settled the mold is placed back into the vacuum chamber and degassed until no more bubbles emerge. The silicone needs to dry for 2–3 h (depending on the size of the mold) until it has reached a coherent state (Kretzer, Soft Robots Tutorial 2012)

Hard silicone mixture (Fig. 4.138).

Fig. 4.138 The Addition Cure 33 silicone rubber part A is mixed with part B in a 1:1 ratio. After stirring the mixture has to be degassed until no more bubbles appear. The liquid is then poured into a flat mold that has roughly the same size as the 3D print in order to create a 1–2 mm thick sheet, which will form the backside of the actuator. The silicone again has to dry until it reaches its hardened state (Kretzer, Soft Robots Tutorial 2012)

Joining the parts (Figs. 4.139, 4.140, 4.141 and 4.142).

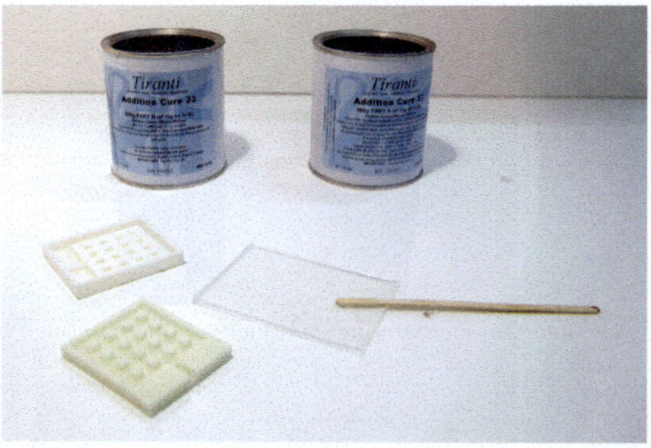

Fig. 4.139 Both, the soft and hard silicone can now be removed form their casts (Kretzer, Soft Robots Tutorial 2012)

Fig. 4.140 Again a small solution of Addition Cure 33 silicone needs to be mixed and degassed. The finished solution is then spread evenly across the clean surface of the hard silicone part (Kretzer, Soft Robots Tutorial 2012)

Fig. 4.141 Using the same mixture a piece of rubber tube is glued into the soft silicone part (Kretzer, Soft Robots Tutorial 2012)

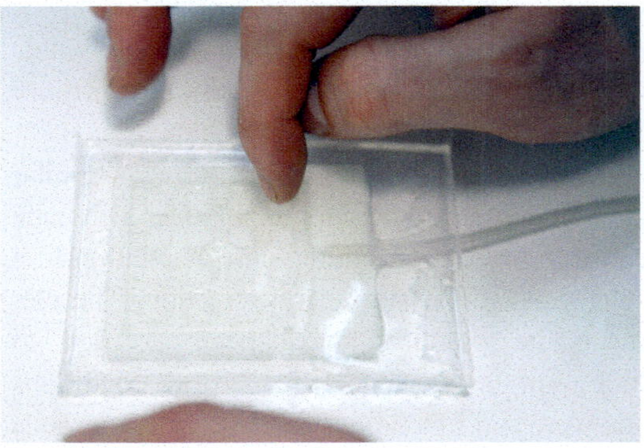

Fig. 4.142 Finally the soft and the hard part are joined together while pressing them firmly. It is essential that the bonding is completely airtight. The component needs to dry for several hours (Kretzer, Soft Robots Tutorial 2012)

Inflation (Fig. 4.143).

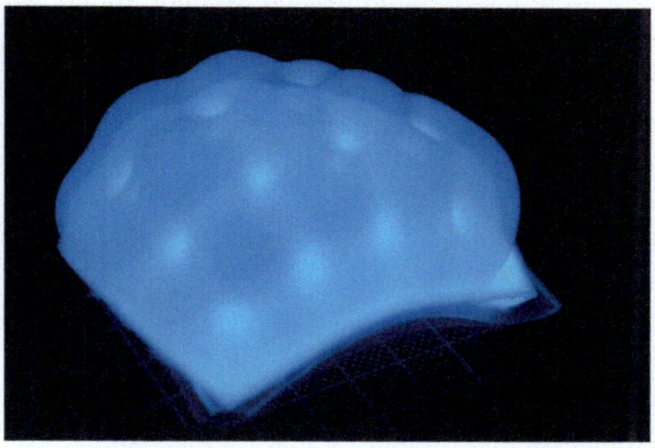

Fig. 4.143 The dry device can now be inflated with a bicycle pump and the correct tube adapter. In order to see the 'glow in the dark' effect the lights have to be turned off (Kretzer, Soft Robots Tutorial 2012)

4.8.3 Case Studies

4.8.3.1 Soft Motion and Liquids (2012)

Soft Motion and Liquids was a three-day workshop held as part of the *Liquid Things* research project at the University of Applied Arts (Angewandte), Vienna from Nov. 30th to Dec. 3rd 2012. The workshop was organized by Roman Kirschner and held by Ben Haworth with additional contributions from Ruairi Glynn and Zoi Karagiozi. It followed a hands-on approach into fabrication methods and actuation of basic soft robotic systems.

The workshop participants worked on a number of previously prepared shapes and forms (Fig. 4.144) and experimented with modifying the material structure. On top of mediating a general understanding of the fabrication process and the working principles of such systems, then in total 10 participants were encouraged to develop their own designs and forge a personal opinion for potential (architectural) applications. Subjects that were explored throughout the workshop considered the scalability and robustness of the devices in order to become architecturally relevant. Moreover, by adding phosphorous powder to the silicone material, devices that would luminescence during darkness were tested, and finally the use of liquids as an alternative to compressed air was evaluated.

Fig. 4.144 One of the many shapes that were created during the Liquid Things Workshop (Kretzer, Soft Robots Tutorial 2012)

4.8.4 Applications

Most current research into soft robotic systems or devices is performed at the Whitesides Research Group and the Wyss Institute for Biologically Inspired Engineering, Harvard University, and is largely funded by the Defence Advanced Research Projects Agency (DARPA) under its maximum mobility and manipulation program. The military interest in the technology is based on developing a robotic system that remains cost effective. Due to the robot's specific resilience it can maneuver through rough terrain and constrained spaces without compromising its shape or structure. On top of potential tactical applications, soft robots could be used in medical systems, stimulating fluid vessels, and muscle movement for modeling or training, or find value in prosthetic or exoskeleton technology (DARPA 2012).

The simplicity and resilience of the systems also imply various possibilities for architecture and design. Especially the nonlinear movement, the organic appearance, and soft surface could allow the creation of adaptive environments that would feel much less mechanic and more natural than current explorations do. However, what remains to be solved will be the massive infrastructure and pneumatic system, needed to actuate the components, when reaching spatial dimensions.

In an attempt to overcome the rigidity, high-energy consumption, and complex control of hard kinetic architectures, Zoi Karagkiozi and Theodoros Themistocleous developed an auxetic pneumatically actuated morphing system. The structure consisted of a number of individual chambers, cross-connected by flexible gateways, and able to deflect in two directions when inflated (Glynn et al. 2013, p. 424). As part of the same research they also developed a kinetic system based on a number of hexagonal inflatable silicone rubber cells. The single-channel elements consisted of a soft silicone layer, which expanded and a hard one, which constrained the soft layer and led to bending motion. By alternating the soft and hard silicone layer, positive and negative curvature could be achieved (Glynn et al. 2013, p. 435).

Daekwon Park at Harvard Graduate School of Design is developing a Soft Modular Pneumatic System (SMPS) which attempts to amplify characteristics of biological design, like modularity, ruggedness, homeostasis, and adaptivity with soft actuator technologies to create a modular, kinetic facade system. Each element in the system is independent from its neighbors through integrated sensors, actuators, and control elements and through this he imagines the emergence of a global behavior due to chain reactions between individual modules at the local level (Meta Territory Studio).

Dino Rossi at ETH Zurich follows a similar attempt but uses the motion of the pneumatic actuator to orient building integrated thin-film solar cells for in situ energy creation. The elements he produced have been exposed to long-term testing under real-world conditions and were still functioning after several thousands of cycles. The successful integration of such a technology at a larger scale would not only reduce mechanics and electronics but also improve performance under harsh conditions and lower production and maintenance-related expenses (Rossi et al. 2013, pp. 0–33).

4.9 Thermochromics

Thermochromic materials belong to the group of chromogenics, which are materials that share the inherent ability to reversibly adjust their color due to an induction process caused by external stimuli. They include electrically switchable technologies (electrochromics), photochromic materials, which can change their hues when exposed to ultraviolet radiation, piezochromics, and mechanochromics, which adjust color when exposed to pressure or stress, and halochromic materials, which respond when the ambient acidity changes. Thermochromics shift their color when experiencing a change in temperature, as displayed in Fig. 4.145 (Lampert 2004, p. 28). The thermochromic color change is based on the reorganization of the molecular or crystal structure of the material when activated. The effect is generally quite noticeable, often dramatic, and happens over a small or distinct temperature interval (Day 1968, p. 649). Thermochromics are used in widely ranging industries and product applications. The most common types are based on either leuco dyes or thermochromic liquid crystals technology and are available at differing activation temperatures.

Fig. 4.145 A disk of thermochromic plastic exposing a *gradual* color change from its cold to warm side (Kretzer, Thermochromic Disk 2013)

- Leuco dyes are usually microencapsulated and exist as dispersant (ink concentrate, pigment), powder, or plastic pellets. They transform from a predefined base color to the colorless leuco form (+5 °C) when heated. This effect is often used in combination with other pigments, which are revealed once the leuco dye disappears. They can also be mixed with further leuco pigments resulting in a gradual change from one color to the next at varying temperatures. Since their temperature response is less accurate than liquid crystals they are generally used for indicators of approximate temperature or various novelty products.
- Liquid crystals are used in dyes and inks and are often microencapsulated. They change color over a much smaller temperature range (+1 °C) than leuco dyes and shift from their clear (black) base color to about every color of the visible spectrum back to clear (black). The effect is based on selective reflection of certain wavelengths when the material's structure changes from its crystalline to the liquid phase. They are used in precision applications, are much more delicate to handle, and require specialized equipment. Also their fabrication is more difficult and hence more expensive than alternative technologies.

4.9.1 History

While almost all inorganic compounds exhibit thermochromic properties to some extent only a few demonstrate significant changes in color. The first systematic investigation of thermochromism dates back to 1871 and was reported by Edwin J. Houston, who heated and cooled a number of solids on copper strips and recorded the ones whose color change was completely reversible. Titanium dioxide (TiO_2) and zinc oxide (ZnO) undergo a (not entirely reversible) subtle transformation, changing from white at ambient temperature to yellow when heated, which is linked to changes in the material's electronic properties (energy levels). More drastic effects can be witnessed in materials that undergo phase transition or display charge-transfer bands near the visible region. This effect is visible, for example, in cuprous mercury iodide (Cu_2HgI_4), which is bright red at low and dark brown at high temperature, nickel sulfate ($NiSO_4(H_2O)_6$), being green at room temperature and yellow at 155 °C, or silver mercury iodide (Ag_2HgI_4) that changes from yellow to orange above 47–51 °C (Day 1968, p. 649).

In 1909, the Austrian chemist Hans Leopold Meyer observed thermochromic reactions in a number of organic compounds. However, an explanation of the phenomenon could only be provided when E. Harnik et al. published a series of articles on the crystal "structure of overcrowded aromatic compounds" in the *Journal of the Chemical Society* in 1954 (p. 3288) and consecutively when J.F.D. Mills and S.C. Nyburg presented their work on "Thermochromism and related effects in Bixanthenylidenes and Bianthronylidenes" in 1963 (p. 308). The first commercial products utilizing thermochromic effects, began emerging from the laboratories in the late 1960s, around the time of the first digital watches, and were mostly based on the same display technology, liquid crystals. Popular gadgets in the 1970s were so-called mood rings, which would change their color spectrum depending on a person's body temperature.

In 1985 G.V. Jorgenson and J.C. Lee published a scientific paper investigating thermochromic materials for optical switching films and their potential solar energy applications in buildings (p. 2). In 1991 a company called Hypercolor began producing thermochromic clothing, especially T-Shirts, based on leuco dye technology that switched in between two colors. In 2003, the Fraunhofer Institute for Applied Polymer Research in Germany successfully developed microencapsulated thermochromic compounds embedded in various plastic composites (Ritter 2007, p. 80). In 2010, the same research group reviewed a range of thermotropic and thermochromic polymer-based materials and their potential as adaptive sun protection. Hydrogel systems, which can act as thermochromic and thermotropic systems, combine excellent light transmittance when off with outstanding solar protection when on (Seeboth et al. 2010, p. 5162).

In 2011, Lars-Olov Hennerdal and Magnus Berggren proposed a full-color thermochromic paper display that would switch in between two images (2011, p. 183303-1). As of now thermochromic materials have always contained at least one toxic component, making them difficult to use in medical or food technology products, but in January 2013 Arno Seeboth, Detlef Lötzsch, and Ralf Ruhmann created the first nontoxic thermochromic polymer based on the biopolymer polylactic acid (PLA) (p. 43).

4.9.2 Material Structure and Operation

The mechanism of the thermochromic reaction varies depending on the material's molecular or crystal structure.

4.9.2.1 Leuco Dye Systems

Thermochromic leuco dye systems perform a reversible change from colored to colorless (translucent) trough molecular rearrangement. Figure 4.146 displays the basic structure and operating principle of a leuco dye system. The microencapsulated particles (1) are usually below 20 μm in diameter and contain a colorant (2), an (organic) acid (3), which is responsible for the reversibility and intensity of the color change, and a low melting solvent (4), whose melting point controls the temperature at which the color change will happen. As long as the temperature is lower than the solvent's melting point the color-forming components are in contact and a visible color appears. When the temperature is increased beyond the melting point of the solvent, the color-forming components are separated and the color disappears. The temperature affects the equilibrium point of the acid/base reaction, which affects the form of the leuco dye. So strictly speaking these systems are not thermochromic but rather halochromic since they change color with a change in pH (Kulcar 2010, p. 272).

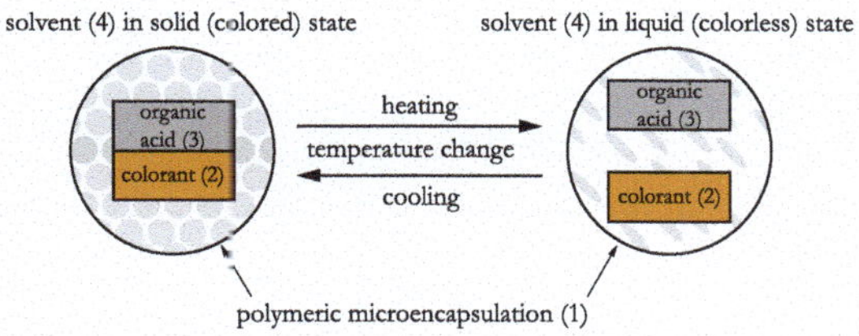

Fig. 4.146 Basic principle of thermochromic leuco dye reaction (adapted with kind permission by LCR Hallcrest Ltd. www.lcrhallcrest.com)

Both the color and the active temperature of the dyes can be controlled and programmed. Commercially, available products exist in a variety of predefined colors and have standard transition temperatures at 16–18 °C, 30–33 °C, 43–45 °C, and 62–65 °C (both base color and temperature may vary depending on the manufacturer). Custom made dyes can be made with temperature thresholds between −15 and −65 °C. The color change takes place over a temperature difference of roughly 5 °C. Due to thermal hysteresis the colors, however already begin to fade at around 4 °C below the actual activation temperature (Ogrodnik 2008, p. 151). The consecutive cooling process hence requires a significantly lower temperature to return to full-color strength (Fig. 4.147).

The dye systems can be further mixed with traditional color pigments, which creates a transition from one color to the other, or blended with other leuco dyes at varying temperatures, allowing a gradual transition through a range of colors. Leuco dyes are available in various forms, including dispersion (ink concentrate or pigment based), powder, or plastic pellets. They can be screen-printed, spray-coated, dispersed, and mixed with polymers or used in bath dyeing systems. Leuco dyes have fairly robust color possibilities and are substantially cheaper and easier to apply and process than other systems. Long exposure to ultraviolet radiation, solvents, or high temperatures (above 200 °C) can lead to a reduced lifespan and irreversible damage.

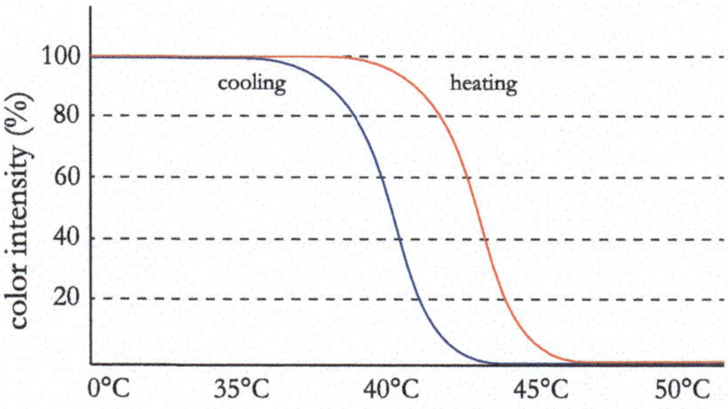

Fig. 4.147 Thermal hysteresis of leuco dye with activation temperature at 47 °C (adapted with kind permission by LCR Hallcrest Ltd. www.lcrhallcrest.com)

4.9.2.2 Liquid Crystal Systems

Liquid crystals are substances in an intermediate state between a crystalline solid and an isotropic liquid. They can be split into two major categories, lyotropic liquid crystals, which result from the action of a solvent, and thermotropic liquid crystals, which emerge from the melting of a solid. Based on their optical properties, thermotropic liquid crystals can be further divided into three distinct types, smectic, nematic, and cholesteric. Cholesteric (also called chiral or twisted nematic) liquid crystals are extremely optically active and well suited for thermochromic applications. Passing through the material, each plane molecular layer (2) is slightly twisted to the one below, resulting in an overall helicoidal structure (1), as displayed in Fig. 4.148 (Hallcrest 1991, p. 23). The pitch (3) defines the repeat distance along the direction of periodicity, which means that layers separated by the pitch exhibit the same molecular orientation (4). Since standard values of the pitch are similar to the wavelength of visible light (400–700 nm), **Bragg diffraction** occurs at the respective layers and the color of the liquid crystals will match the wavelength of light for which highest constructive interference appears (White and LeBlanc 1999, p. 1201).

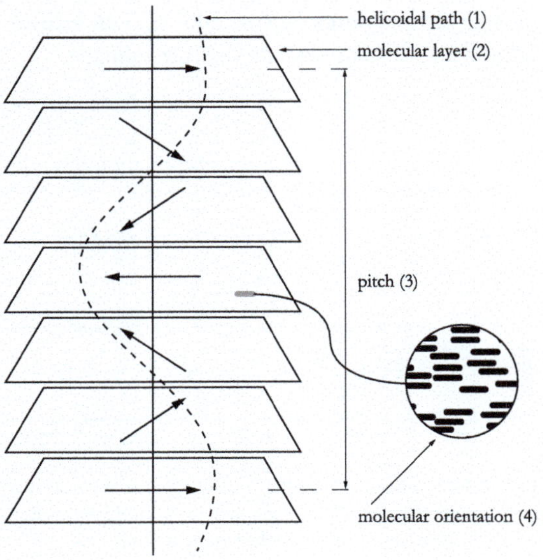

Fig. 4.148 Molecular orientation in cholesteric liquid crystal structure when passing through the material (adapted with kind permission by LCR Hallcrest Ltd. www.lcrhallcrest.com)

In a cholesteric liquid crystal, changes in temperature will lead to thermal expansion and hence a change in layer spacing and pitch. As a result the material reflects different colors when transforming from the low-temperature crystalline phase through the cholesteric phase to the high-temperature isotropic liquid phase; an effect also referred to as color play. While the material is generally clear and nonreflective, it is usually applied to a black background to reduce loss of color due to selective reflection. The color then moves from black to red (at low temperature) through orange, yellow, green, blue, and violet (at high temperature) back to black as shown in Fig. 4.149 (Stasiek and Kowalewski 2002, p. 1). Thermochromic liquid crystals are generally microencapsulated (in the range of 10–15 μm in diameter) and dispersed in a polyvinyl alcohol binder, resulting in slurry, which can be screen-printed, sprayed, or coated onto a variety of surfaces. Thermochromic liquid crystals can exhibit color change over a wide range of temperatures from as low as −30 °C to above 120 °C. They are highly accurate and used in applications where the color change has to be precisely defined. However, they are quite difficult to work with and require specialized equipment. Also the material is more expensive than alternative technologies and UV radiation, chemicals, or solvent can reduce its lifespan.

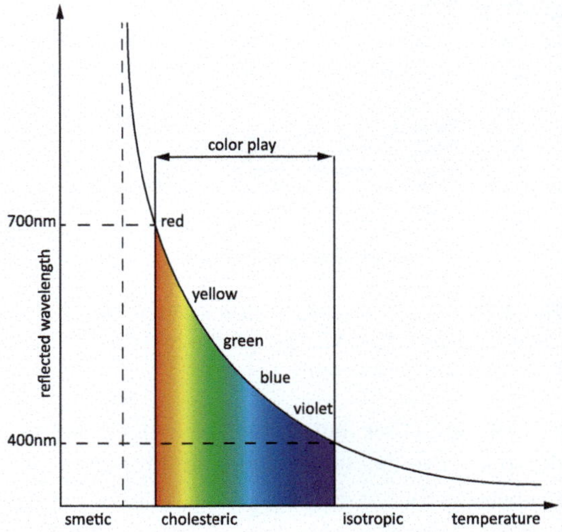

Fig. 4.149 Typical reflected wavelength (*color*) temperature response of thermochromic liquid crystal mixture (adapted with kind permission from Springer Science +Business Media from Stasiek and Kowalewski 2002, Fig. 1)

4.9.3 Fabrication

Both thermochromic leuco dye and liquid crystals systems are available as base substrates and microencapsulated. Microencapsulation helps making them more durable and enhances their spectrum of applications. Both types come in various forms, including inks, pigments, or pellets and can be processed using different techniques. Extensive research has been performed in their use on fabrics in combination with embedded electronics for the creation of smart textiles.

In the following tutorial leuco dye pigments are mixed with polyester resin to produce a color-changing plastic. Polyester resin is a thermosetting resin produced by a reaction between several organic acids and polyhydric alcohols. Industrially, used for molding reinforced fiber and composite products, polyester resin was chosen primarily for its perfect binding of the thermochromic pigments. In addition, its property to take on almost any form and texture, its aesthetic qualities of translucency, surface quality, and relative ease of post-cast processing made it an ideal choice of material.

4.9.3.1 Materials and Tools

See Fig. 4.150.

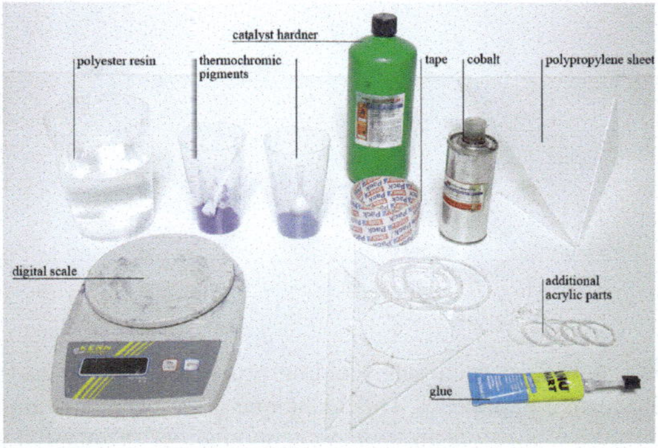

Fig. 4.150 Polyester resin/thermochromic leuco dye pigments/catalyst hardener/cobalt/polypropylene sheet/ scale, tape, acrylic pieces, glue (Xydis and Shammas, Thermochromics Tutorial Pictures 2013)

4.9.3.2 Steps

Preparation (Fig. 4.151).

This tutorial focuses solely on the creation of a color-changing plastic container. It does not cover the integration of additional electronic components that will be needed in order to heat and cool the material in a controlled manner, as shown in Fig. 4.152. The procedure should be executed in a well-ventilated space and respiratory as well as eye protection and gloves are encouraged during all stages due to toxic fumes that might emerge throughout the process. The ambient temperature should be around 20 °C. Working in an environment below 13 °C is not recommended since curing might not start and the results will remain unsatisfactory.

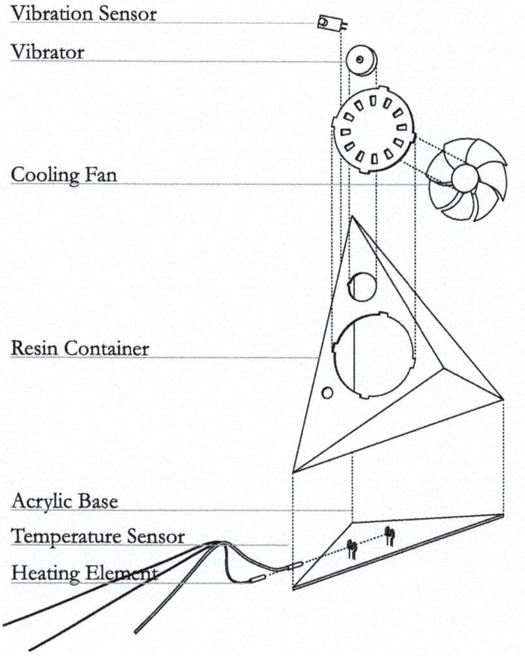

Fig. 4.151 Explosion diagram of hollow thermochromic component including additional electronic parts, which are not part of the tutorial (Xydis and Shammas, Thermochromics Tutorial Pictures 2013)

Shape design (Fig. 4.152).

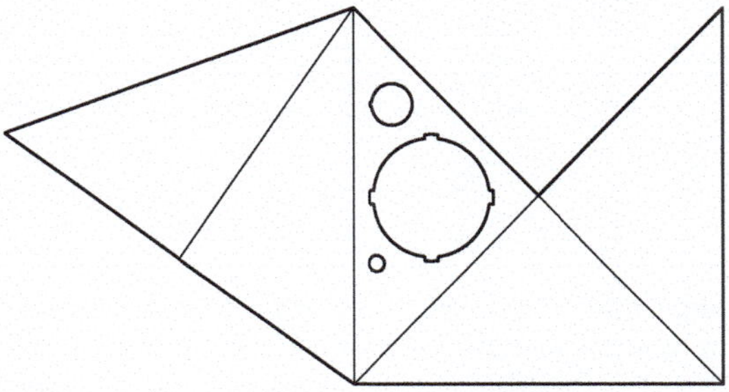

Fig. 4.152 The shape can be designed on a computer using a graphical program. It needs to be developable so it can be cut from flat sheets. The shapes should be cut from 2-mm flexible polypropylene sheets using a laser cutter (Xydis and Shammas, Thermochromics Tutorial Pictures 2013)

Gluing acrylic parts (Fig. 4.153).

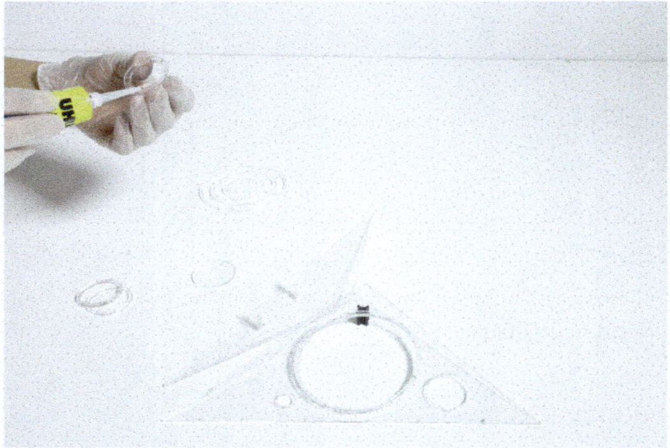

Fig. 4.153 In the particular shape of this project two out of six sides were made from acrylic plates that would later house and support the supplementary electronic components. The clear acrylic pieces need to be glued together according to the design (Xydis and Shammas, Thermochromics Tutorial Pictures 2013)

Mold assembly (Fig. 4.154).

Fig. 4.154 The polypropylene and acrylic parts need to be joined into a three-dimensional object using scotch tape. It is important to avoid wrinkles in the tape since these might lead to leakage later. If the mold is curved, small pieces of tape will help to interpolate the curve and prevent wrinkles. All holes of the mold have to be carefully closed with tape, except for one opening that is needed to pour in the resin (Xydis and Shammas, Thermochromics Tutorial Pictures 2013)

Thermochromic resin mixture (Figs. 4.155, 4.156 and 4.157).

Fig. 4.155 A rotational molding process is used to cast the resin components. This technique allows the creation of hollow components with an economic use of material. Additionally it creates a visually interesting effect of layer variation within each component. The mixing ratios are: 100 ml (110 g) of resin with 4 g of leuco dye pigment, 2.5 ml of hardener and one drop of cobalt (Xydis and Shammas, Thermochromics Tutorial Pictures 2013)

Fig. 4.156 A larger amount of pigment will lead to a faster color reaction. Since various types of polyester resin and hardener exist specified mixing ratios by the manufacturer should be used. If unclear the material supplier has to be contacted. The components in this tutorial have an area of almost 100 cm^2 and need a wall thickness of 3 mm; hence 300 ml (350 g) of resin are used per piece. Two different colors are added for each component, a *blue* base color (6 g) which turns translucent at 33 °C and a *purple* color (6 g) dissolving at 44 °C. Both, the resin and the pigments are mixed and stirred until no clumps remain (Xydis and Shammas, Thermochromics Tutorial Pictures 2013)

Fig. 4.157 Once they've completely dissolved 7.5 ml of hardener is added as well as a drop of cobalt. The cobalt increases the speed of the curing process and needs to be handled with extreme care, since excess could raise the temperature to dangerous degrees. The liquid is then stirred until all parts are evenly dissolved (Xydis and Shammas, Thermochromics Tutorial Pictures 2013)

Shape fabrication (Figs. 4.158, 4.159 and 4.160).

Fig. 4.158 Now the thermochromic resin mixture is poured into the mold (Xydis and Shammas, Thermochromics Tutorial Pictures 2013)

Fig. 4.159 Once everything is inside the mold the opening needs to be closed with scotch tape (Xydis and Shammas, Thermochromics Tutorial Pictures 2013)

Fig. 4.160 The mold is then continuously rotated around all axes. The process should be at consistent speed and motions be repeated to ensure an even spread of material. After 10 min the resin begins producing heat and turns extremely liquid, almost watery as well as slightly change its color. Despite this effect, it is important to keep rotating exactly as before. After another 10 min it becomes more viscous again. This marks the most important state, and rotation should be continued with extra care, before the mixture solidifies 1–2 min later. The piece then has to rest for about 5 min before doing the next step (Xydis and Shammas, Thermochromics Tutorial Pictures 2013)

The openings (Fig. 4.161).

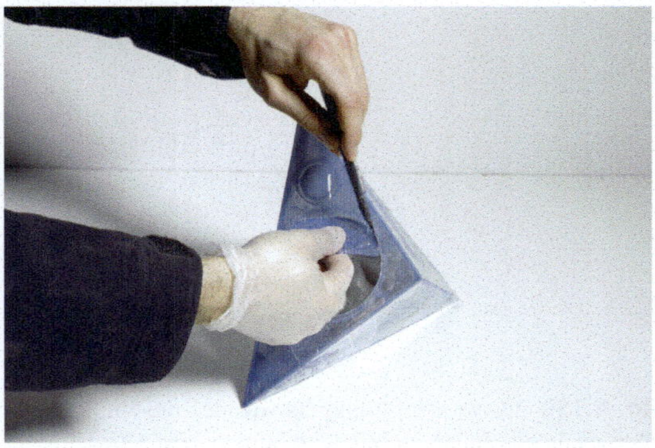

Fig. 4.161 Since at this stage the resin is already solid but not completely hard, the necessary openings can be cut with a utility knife. The openings need to be cleaned thoroughly. The components then have to cure for at least 12–24 h, depending on the shape and thickness of the material (Xydis and Shammas, Thermochromics Tutorial Pictures 2013)

Unwrapping (Figs. 4.162 and 4.163).

Fig. 4.162 With a knife cutter and cautious force the polypropylene sheets have to be cut open on the edges (Xydis and Shammas, Thermochromics Tutorial Pictures 2013)

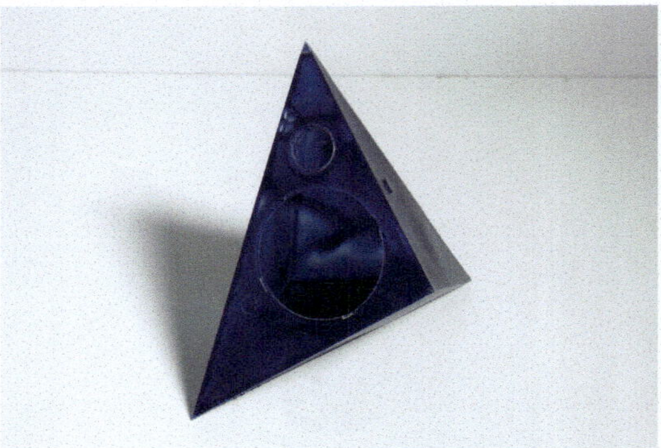

Fig. 4.163 Once the final plastic shape has been unwrapped the knife can be used to clean the corners (Xydis and Shammas, Thermochromics Tutorial Pictures 2013)

4.9.4 Case Studies

The properties of thermochromic plastic objects have been explored in two consecutive projects at the Chair for CAAD, ETH Zurich as well as a one-week workshop at the Institute of Advanced Architecture of Catalonia. The first project, a touch responsive installation called *Resinance*, was realized during 5 weeks as part of a MAS CAAD module in spring 2013. The second project, *Resinance 2.0*, was building upon the technology and experience of the first. It was realized and exhibited at the *Acadia 2013 Conference* in Cambridge, Ontario, Canada in October 2013. The workshop *Dynamics in Extreme Environments* took place in January 2015 and was a collaboration between the Chair for CAAD, ETH Zurich, IaaC, and Chalmers University of Technology.

4.9.4.1 Resinance (2013)

The design of *Resinance* (Fig. 4.164) was inspired by the behavior of simple organic life forms, in specific the formation of cellular colonies. It represented an ecology of functional units that could both work independently but also in unison with their neighbors, consisting of 40 active elements that were slowly changing their surface color in response to tactile input. While this gradual transformation as such could not directly be observed, each device incorporated a secondary actuator, providing instant response through vibrations. Every four elements were joined by a control node that formally resembled the rest of the units but without the ability of changing color. The nodes both directed the behavior of the cluster and transferred the present state of each piece to its neighbors. Therefore, the physical input not only altered the affected element but was also transmitted throughout the complete network in swarm like manner.

Fig. 4.164 The elements of Resinance changed their surface color when touched physically (Xydis and Shammas, Resinance 2013)

The four-week course was split into three parts. During the first part the students explored the design and fabrication of three-dimensional hollow modules produced from a polyester resin that had been augmented with thermochromic pigments (Fig 4.165). The second part focused on the sensing and transforming capabilities of the units and the integrated electronics. The last part considered the communication between the different elements, their emergent behavior, and networked agency.

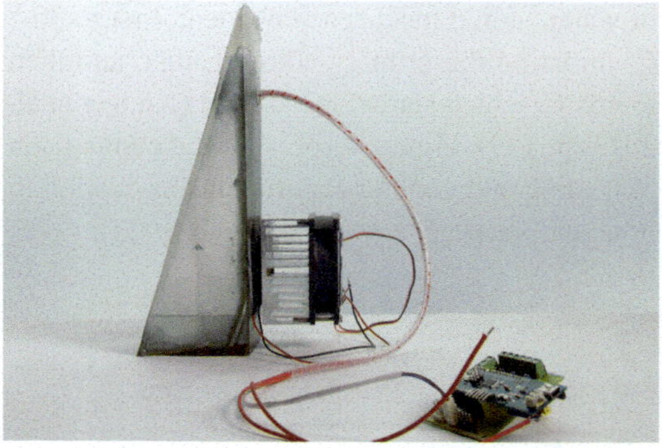

Fig. 4.165 Early prototype developed during the course (Xydis and Shammas, Resinance 2013)

The crystal-shaped elements of the installation were produced through a rotational molding technique. After a number of design experiments, it was concluded that a simple form would prove most effective in demonstrating the thermochromic properties of the resin. The faceted geometry of the elements offered a strong contrast to the organic fluidity of the material behavior.

A transparent control unit grouped every four components in the system. These in total 10 control nodes contained an **Arduino Fio**, a connection to a 12 V power supply, an **XBee** Series 1 Wi-Fi component, and a custom-built Arduino shield that guided signals between the arrays of sensors and actuators. Every component had a piezo-vibration sensor acting as a touch sensor, a thermistor as internal temperature sensor, a heating element, a modified CPU fan for cooling, and a vibrating motor. Due to the very tactile properties of the plastic material, in conjunction with the change in temperature, it was essential that the primary interaction would focus on touch. When a component was touched, it responded with strong 'shivers' as a form of direct haptic feedback. The length and strength of the vibration resulted from the number of times the component had been touched, the intensity of the touch, as well as the current state of its neighbors. On top of the shivering, the touch input triggered a second, much subtler reaction: the thermochromic color transformation. This was accomplished by heating 200 ml of water, stored inside each element, first to 30 ° C, initiating the primary color change, and then to 46 °C, which started the second (Kretzer et al. 2013, p. 139). Figure 4.166 shows a comparison of heating and cooling experiments in relation to time and shape.

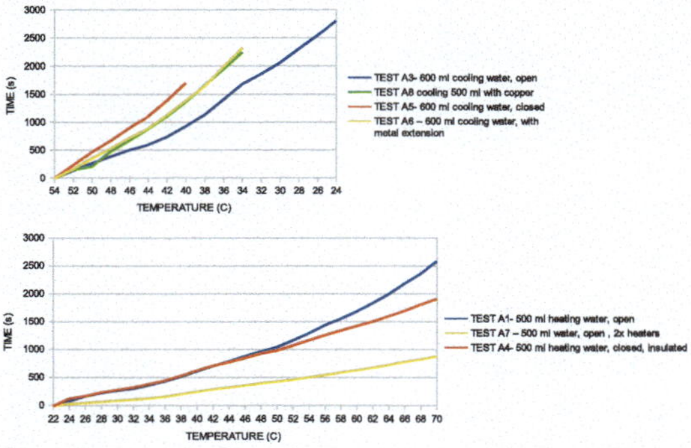

Fig. 4.166 Comparison of time needed to cool and heat various shapes with different amounts of water using various techniques (Smigielska, Time/heat comparison 2013)

The heating element and thermistor were suspended in the water, which worked both as a conductor for thermal energy on the lower portion of the element, and as a source of steam which heated the higher portion of the component. The color change began within the first minute after activation, and reached its peak after 10 min. A fan then vented out the hot air, so that the color shifted back in 2–4 min (Figs. 4.167, 4.168 and 4.169).

Fig. 4.167 Single component during thermochromic color transformation from *purple* to *red* to clear (Xydis and Shammas, Color Change 2013)

Fig. 4.168 Single component during thermochromic color transformation from *purple* to *red* to clear (Xydis and Shammas, Color Change 2013)

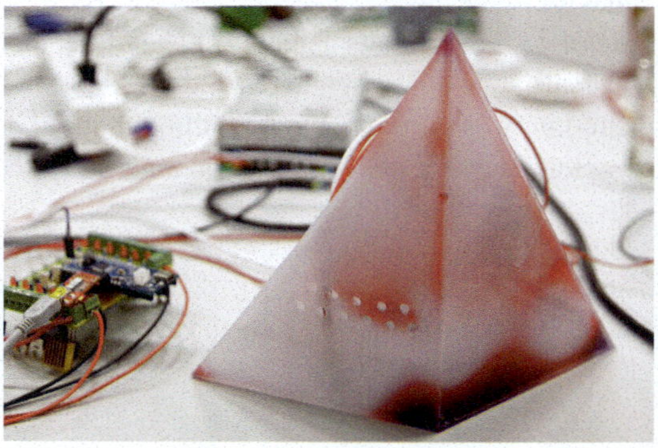

Fig. 4.169 Single component during thermochromic color transformation from *purple* to *red* to clear (Xydis and Shammas, Color Change 2013)

To join all parts into a consistent installation a table-level design allowed for direct interactivity and easy configuration of electronics by providing space for cables and power supplies underneath the platform. The geometry of the plinth was based on a Penrose tessellation matrix and was fabricated on a CNC milling machine from polystyrene foam boards and painted with epoxy resin (Fig. 4.170).

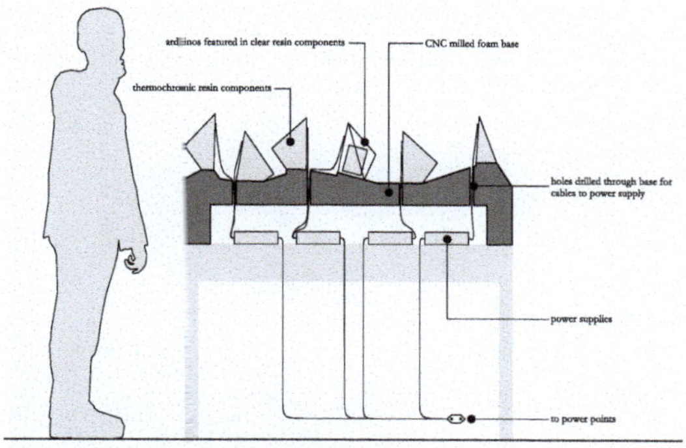

Fig. 4.170 Section through the installation displaying the variety of active elements, electronic infrastructure, and base platform (In, Resinance 2013)

The control system of the installation involved programming and networking as a distributed network. To realize this approach, each cluster had its locally operating control unit, which was configured so that messages were sent only to and from its closest neighboring clusters. The global and emergent behavior of the entire installation was decomposed by the software into individual, simple responses, so that all necessary computation was ultimately performed by the network. In this way, the network of devices became close to an ecosystem of artificial organisms, locally informing and influencing one another (Fig. 4.171).

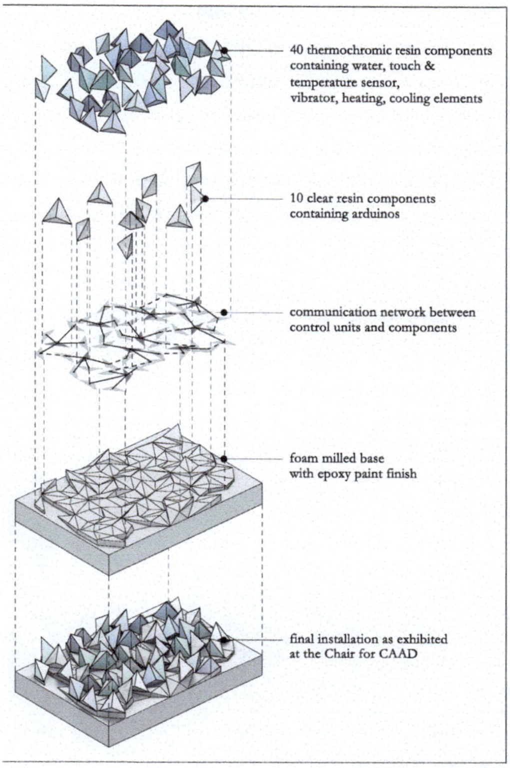

Fig. 4.171 Explosion diagram of various layers of installation. From *top* to *bottom*: thermochromic containers, cluster controllers, communication network, foam base, final installation (In, Resinance 2013)

As a result, the created network was highly stable. Potential failures of single nodes did not obstruct the global behavior. Moreover, the reprogramming of the installation could be done virally, which made modifications, adjustments, and extensions trivial to perform. A new group of elements could simply be connected to the others and receive the behavior-defining code from the adjacent elements. The finished installation was set up and exhibited for several weeks at the Chair for CAAD, ETH Zurich (Fig. 4.172).

Fig. 4.172 Final installation during its color transformation (Xydis and Shammas, Resinance 2013)

4.9.4.2 Resinance 2.0 (2013)

Resinance 2.0 was the successor of *Resinance*, realized 6 month after the first project. Its general system was building upon the initial installation with improved behavioral complexity and technical and material resilience. The project emerged from a student application to showcase the work at the *2013 ACADIA Conference* at the school of architecture, University of Waterloo, Cambridge, Canada (Fig. 4.173).

While the main concept of the installation was similar to the previous one (responsive smart material elements that would change color when physically touched and share the information with their neighboring elements in order to develop a global emergent behavior based on local interactions), several parts of the installation were significantly different (Kretzer 2014, pp. 468–469).

Fig. 4.1.173 Resinance 2.0 as displayed at the ACADIA 2013 conference, school of architecture, University of Waterloo, Canada from Oct. 24–26, 2013 [Xydis, Resinance 2, 2013 (here is also a mistake in the numbering: 4.1.173!!)]

Topology

The layout of the installation was changed to a linear arrangement consisting of 10 clusters, each containing three elements (Fig. 4.174). The clusters were linked wirelessly and constantly communicated their current state to a master node, which compiled the information and gave it back to the respective module. Every cluster was sitting atop an acrylic base that provided stability when the elements were moving and housed the necessary electronic and mechanical components.

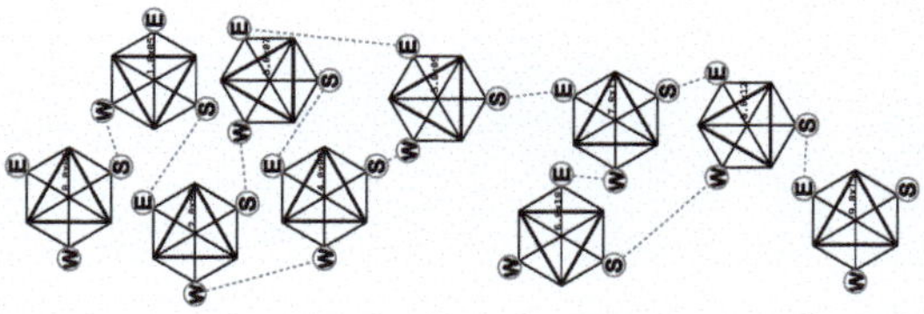

Fig. 4.174 Topological layout (*top-view*) of Resinance 2.0 (Xydis and Shammas, Resinance 2 2013)

Sensing

The sensing capabilities of the individual elements had been vastly simplified. While in *Resinance* piezo-vibration sensors were used to measure human interaction, in *Resinance 2.0* a metallic mesh had been embedded into the polyester resin walls. The mesh, which was added during the rotational casting process, was used as a capacitive proximity sensor. This allowed visitors to interact with the modules all over their surface and directly experience the change of temperature (Fig. 4.175).

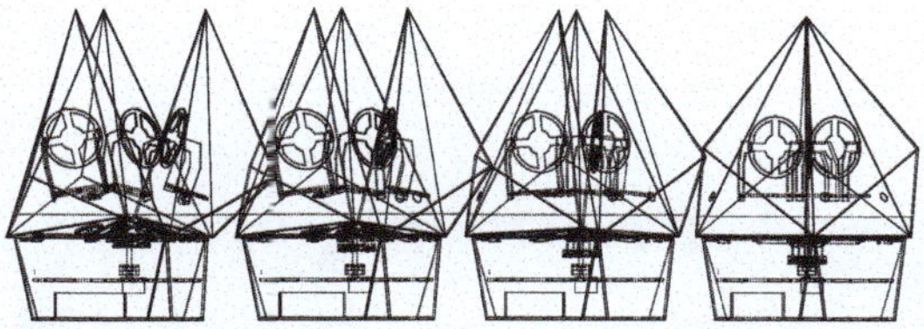

Fig. 4.175 Section through clusters and modules in various states (Xydis and Shammas, Resinance 2 2013)

Actuation

In addition to the shivering motion, which was improved from the previous setup by moving the vibrators from the side to a less obtrusive position at the bottom of the individual elements, a stepper motor was included, which slowly raised the center of each cluster when the elements had reached their peak temperature. Since the fans were facing inwards, due to aesthetic reasons and in order to speed up the heating process, the motion, which resembled a blossoming flower, opened the air inlets and was hence triggered simultaneously with the cooling process. Moreover, the color change was drastically improved up by using two heaters per element instead of one. An overview off all involved components is given in Fig. 4.176.

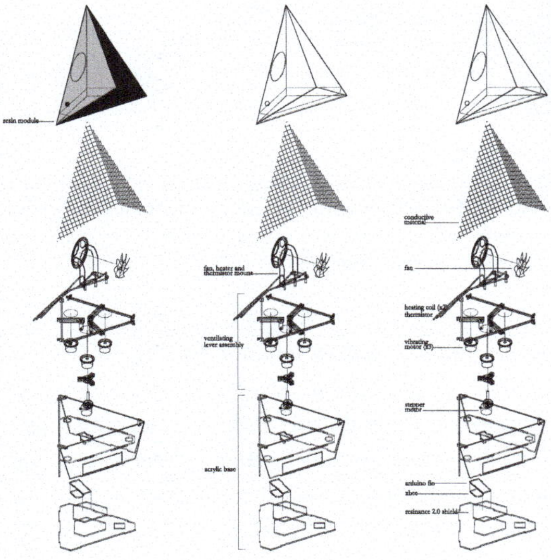

Fig. 4.176 Explosion diagram of individual component; *left* resin module; *center* support structure; *right* electronic parts (Xydis and Shammas, Resinance 2 2013)

Communication

The exchange of information between the individual clusters happened wirelessly. Custom designed shields were used to control the various electronic components, like heaters, temperature sensors, vibrators, motors, etc. The shields were attached to Arduino Fio boards equipped with X-Bee radios in order to communicate with adjacent clusters (Fig. 4.177). A visual interface, which ran on a nearby screen, graphically displayed every element, their current temperature, whether they were in a heating or cooling process, how often they had been touched and when the last touch had occurred. The elements had not only memory of their popularity, measured by the amount of touches, but also tried to return to their initial calm state by gradually deducting points from their counter if not enough interaction would happen. The duration of this process was largely depending on the popularity of the respective element.

Fig. 4.177 Resinance 2.0 during different stages of its metamorphosis (Xydis, Resinance 2013)

4.9.4.3 Dynamics in Extreme Environments (2015)

Dynamics in Extreme Environments was a four-day workshop held at the Institute of Advanced Architecture of Catalonia (IaaC) in January 2015. The workshop was supervised by Areti Markopoulou (IaaC), Alexandre Dubor (IaaC), Carlos Bausa (IaaC), Manuel Kretzer (ETH CAAD), and Stig Anton Nielsen (Chalmers University). It departed from two emerging architectural tendencies. First, it highlighted a growing interest in the creation of responsive spatial environments, fostered through novel material developments and technologies. Such materials challenge the predominant notion of static architectural forms and conversely open up peerless possibilities for the development of adaptive architectural models free from mechanical complexity. Second, the workshop wanted to address a number of rising twenty first century challenges largely related to issues of global warming, such as shifts in temperature and weather, which not only affect the ecological environment but also raise questions regarding future urban settlements and new, appropriate ways of building and inhabiting. During the course the students were introduced to a specific selection of (temperature) responsive materials and related fabrication techniques. The thirty students, who were split into five groups, gained insights in thinking in terms of behaviors and dynamic properties. Based on this they were asked to approach the task from two directions, firstly an unconstrained material exploration, elaborating upon particular phenomena, and secondly a focus on human situations that would benefit from spatial adaptability (Fig. 4.178).

Fig. 4.178 IaaC students and faculty looking at one of the prototypes, which were developed during the four-day workshop (IaaC, Dynamics in Extreme Environments 2015)

The group *Thermoteam* explored the color-changing effect of thermochromics in various patterns for the creation of a dynamic double-layer facade system (Fig. 4.179).

Fig. 4.179 Thermoteam working prototype in scale 1:10 during color transformation (Jun, Dynamics in Extreme Environments 2015)

Survival Bubble was a concept for a floating emergency structure that actively deforms when exposed to sun radiation. The group *Thermoluminescence* developed a color-changing composite material by alternating layers of fabric and thermochromic polycaprolactone. The modules were to be joined into an active membrane that should allow shadow and shelter (Fig. 4.180).

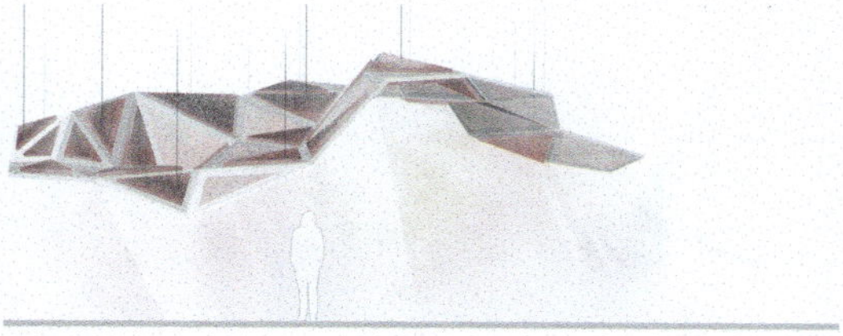

Fig. 4.180 Section of project thermoluminescence through envisioned canopy system (IaaC, Dynamics in Extreme Environments 2015)

The team behind *Delaying Desertification* (Fig. 4.181) produced hollow elements that contained a mixture of sodium bi-carbonate and acetic acid, which when heated lead to a chemical reaction that inflated the units. Once the material had cooled down again the emerging bubbles turned solid and created visually mesmerizing heterogeneous surfaces.

Fig. 4.181 Self-inflating unit of project delaying desertification (IaaC, Dynamics in Extreme Environments 2015)

Maneuvering Reticulation focused on 3D printing flexible and functional components. Some elements incorporated a capillary system that carried warm water to achieve selective color changes. Others included 3D printed air muscles to actuate the complex structure (Fig. 4.182).

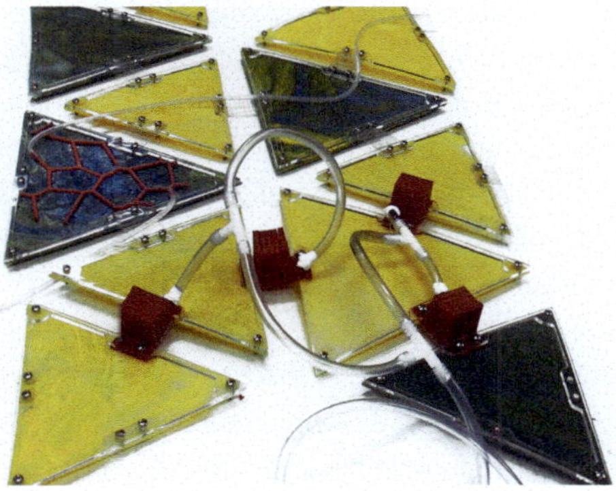

Fig. 4.182 Prototype of maneuvering reticulation showing elements connected through air muscles with tubing and embedded capillary system (IaaC, Dynamics in Extreme Environments 2015)

4.9.5 Applications

Leuco dye thermochromic materials have been used extensively in a large number of novelty and household applications where the color change does not need to precisely respond to the change in temperature. This includes toothbrushes, which switch color after being held for approximately 2 min, several baby items like spoons, cups, or plates that adapt once the warmed up food has reached the desired temperature, food packaging, such as beer cans and bottles, which change their label when cooled, kettles that become transparent when boiling, or battery tester strips, which respond to the electronic resistance in the conductive ink. Very popular in the 1990s were the Hypercolor T-Shirts, however, the sensitivity of thermochromics toward UV radiation and certain chemicals and washing detergents still prevents their larger use in everyday textile applications. The London-based fashion house Cute Circuit has produced experimental clothing, which changes its color and pattern depending on the body temperature of the wearer (Seymour 2009, p. 42). Marjan Kooroshnia, doctoral student at the Swedish School of Textiles, has done extensive research on leuco dye-based thermochromics on fabrics (Kooroshnia 2013, pp. 1–6) and recently proposed a face mask which measures the exhale temperature and indicate if the wearer has a fever.

In 2006, the Japanese artist and educator Akira Wakita created Fabcell, a flexible, non-emissive, multicolor fabric module, in which fibers dyed with liquid crystal thermochromics are interwoven with conductive yarns and connected to electronic components (Metamorphic Architecture). The US-based company Body Faders sells heat activated thermochromic fabrics as part of their Novachromic series (Body Faders).

Liquid crystal thermochromics first became very popular in the 1970s for their use in mood rings, which showed different colors depending on the wearer's body temperature. Today they are mostly used as thermometers in refrigerators, aquariums, and rooms or for medical purposes, as they have considerable advantages over other thermometers, like being flexible, virtually unbreakable and much cheaper to produce. They are also applied in engineering applications, like monitoring the temperature of surfaces, liquids or gas, or the thermal transfer between different materials (Sage 2011, p. 1560). In electronics they are used to detect short or open circuits and hence prevent potential fire hazards (Ogrodnik 2008, pp. 150–155). The online distributor www.

inventables.com offers temperature-sensitive glass tiles, which fade through three color phases, one with each 6–10 °C rise in temperature (Inventables).

In architecture and especially furniture design thermochromics have been used extensively to indicate the thermal trace of a person at a particular time and location. Many of German architect Jürgen Mayer H.'s interior designs, often called Housewarming, are responsive to body temperature and display a colored imprint of people using the space, which fades over time (Ritter 2007, p. 87). Zane Berzina's playful Touch Me installations (2005–2011) invite visitors "to create visual responses through touch and metamorphoses from one state into another" (Berzina). Both Elisa Strozyk and Shi Yuan have developed heat sensitive wallpapers that display floral and ornamental patterns when experiencing a rise in temperature, visualizing the on/off state of radiators or other heating elements (Strozyk). Shi Yuan has applied the same technique to other paper-based objects like posters or calendars (Dornob—design ideas daily). In a paper presented at the 2009 eCAADe conference, David van der Maas et al. describe a number of experimental prototypes, exploring the use of thermochromic ink as a material for the design of architectural information surfaces, with their main goal to reach spatial dimensions (Van der Maas et al. 2009, p. 491).

Recently, liquid crystal thermochromics have gained increasing importance in the production of smart coatings, which can respond to their environment. On the exterior of buildings thermochromic paints are sought to reduce energy consumption, as they would change to white when getting hot. An effect like this could lead to a reduction in the necessity for air conditioning (Miodownik 2008, p. 6). Unfortunately the main challenge remains to deal with continuous exposure to UV light, which will cause the material to degrade and lose its color-changing abilities.

Embedded in a window system thermochromics can be used to change from less light absorbing to more light absorbing as the temperature of the material is increased and by this control the transmission of visible light and the flux of solar energy. When above a certain threshold temperature such a smart window would reflect the infrared portion of the incoming light and hence significantly reduce the amount of heat gain (Parkin et al. 2008, p. 2). The US company RavenBrick commercializes thermochromic Smart Windows since 2006 (RavenBrick).

4.10 Summary and Conclusion

> The limits of my language mean the limits of my world.
> Wittgenstein (1971), Tractatus Logico-Philosophicus, pp. 115–117

The present chapter is trying to provide a broad but comprehensive understanding of the distinct domain of *information materials* by expanding upon a selection of eight material phenomena, *Aerogels*, *Bioluminescence*, *Bioplastics*, *Dye-sensitized Solar Cells*, *Electroluminescent Displays*, *Electroactive Polymers*, *Soft Robotics*, and *Thermochromics*. Each material is described in terms of its historical development and usage, its working principle, step-by-step fabrication, and architectural application. The fabrication procedure and functionality of certain materials has been tested and verified during experimental installations and case studies for their (speculative) usability. The chosen materials are selected based on interest into their particular behaviors, possible usage, or appearance. The collection does not represent a finite assembly but is meant as the beginning of an open-ended overview, empowering the interested reader in becoming literate in the use of *information materials*. It will continue to grow, beyond the scope of this publication, in its online counterpart www.materiability.com and become a useful point of reference and inspiration. The structure and language of the essays is aimed at architects and designers, yet each section has been verified by researchers from the respective domain.[1]

The various material parts were written in the period between 2012 and 2015 with the demand

[1]The researchers are mentioned in the *Credits* (Appendix A.1).

to represent the latest state of development. Yet, due to the quickly changing nature of technological progress the sections are focused more on the descriptions of a material's 'Structure and Operation' or it's 'Fabrication' than the 'History' and especially the 'Applications' parts. Emphasizing these areas instead of actual usages first ensures that a material's functionality does not get too strongly associated with a certain application and second allows the work to remain long-term validity albeit the respective material might in the meantime have evolved further.

The described materials only represent a small fraction of the field of *information materials*. There are many others, which are worth describing and might in the future be added to the online database. The purpose of this chapter is not in suggesting particular usages related to certain materials but rather in unfolding a previously hard-to-reach territory. Therefore, the materials are not sorted according to specific properties but in alphabetical order. Yet, as already mentioned in the introduction to this chapter, they do obviously share myriad similarities. These may include their phenomenological behavior (shape-changing, light-emitting, electricity-creating,…), physical properties (flexibility, weight, opacity, conductivity, solidity, resistance, size,…), fabrication procedure, response stimuli, and activation (thermal, electrical, UV radiation,…), or material class (ceramic, polymer, semiconductor, metal,…).

The decision to not provide a system addressing these categories is due to the small number of selected materials, the format of this publication in print,[2] and especially to encourage the reader in developing their own opinions and ideas. Finding such commonalities, which go beyond the obvious may lead to extremely powerful, unprecedented visions and applications. Yet, many of the materials also have significant deficiencies especially in regards to long-term performance and stability in response to environmental influences, such as UV radiation. In architecture, where materials are ought to last for many years, such issues can quickly prevent them from being applied and therefore should be addressed early enough during a project's development. This requires architects and designers to see these materials not as finished products but to understand them as dynamic elements, which are yet to be developed further. By taking responsibility in giving them architectural relevance architects can become integral parts of their progress, mediating in between related disciplines and communicating certain requirements. Often these materials are, for example, designed for products where immediacy or resolution is key, qualities, which in architectural cases could maybe be traded for a rise in scale or longevity.

Moreover, aspects like temporality or decay could become essential design components, allowing a building to change and maybe even partially die with and in response to its occupants. The notion of such a 'living' architecture will be discussed further in the following and final chapter.

[2]Multidimensional categorization in printed form is difficult and can quickly become confusing.

References

Abileah A, Harkonen K, Pakkala A, Smid G (2008) Transparent electroluminescent (EL) displays (planar systems). http://www.all-electronics.de/download/textmedia/11856/411-pdf-planar-transparent-el-displays-whitepaper-1.pdf. Accessed 15 Oct 2013

Achari MB, Elumalai V, Vlachopoulos N, Safdari M, Gao J, Gardnera JM, Kloo L (2013) A quasi-liquid polymer-based cobalt redox mediator electrolyte for dye-sensitized solar cells. Phys Chem Chem Phys 15 (40):17419–17425

Aerogel.org. The 2000s: new possibilities and commercialization. http://www.Aerogel.org/?p=829. Accessed 8 Nov 2013

American Chemical Society (ACS) (2012) Materials inspired by mother nature: one-pound boat that could float 1,000 pounds. ScienceDaily. http://www.sciencedaily.com/releases/2012/03/120325173010.htm?utm_source=feedburner&utm_medium=feed&utm_campaign=Feed%3A+sciencedaily+%28ScienceDaily%3A+Latest+Science+News%29. Accessed 26 Nov 2013

Avérous L, Pollet E (2012) Biodegradable polymers. In: Avérous L, Pollet E (eds) Environmental silicate nano-biocomposites, green energy and technology. Springer, London, pp 13–39

Bar-Cohen Y (2002) Electro-active polymers: current capabilities and challenges. In: Proceedings of SPIE 4695, smart structures and materials 2002: electroactive polymer actuators and devices (EAPAD). doi:10.1117/12.475159

Bar-Cohen Y (2004) Electroactive polymers history, current status, and infrastructure. In: Bar-Cohen Y (ed) Electroactive polymer (EAP) actuators as artificial muscles—reality, potential, and challenges, 2nd edn. SPIE Press, Bellingham, pp 3–43

Bar-Cohen Y (2005) 1st wrestling match between human and electroactive polymers actuated robotic arms was held on March 7, 2005 in San Diego, CA. In: Worldwide electroactive polymers newsletter, vol 7(1)

Barraud E (2012) Stained glass windows for the Swiss Tech Convention Center. EPFL News Mediacom. http://actu.epfl.ch/news/stained-glass-solar-windows-for-the-swiss-tech-con. Accessed 10 Sept 2013

Barrow WA (1999) Electroluminescent displays. In: Webster JG (ed) The measurement, instrumentation and sensors handbook. CRC Press, Boca Raton, p 96-14

Berzina Z. Touch me (2005–2009). http://www.zaneberzina.com/touchme.htm. Accessed 26 Nov 2013

Body Faders. Novachromic special effects technology. http://www.bodyfaders.com/shopping/pc/viewCategories.asp?idCategory=29. Accessed 26 Nov 2013

Bozic-Weber B, Constable EC, Fürer SO, Housecroft CE, Troxlera LJ, Zampesea JA (2013) Copper(I) dye-sensitized solar cells with [Co(bpy)3]2+/3+ electrolyte. Chem Commun 49(65):7222–7224. doi:10.1039/C3CC44595J

Brunner M, Welker T (2005) Electrical and optical properties of screen-printed electroluminescent structures. In: Proceedings of SPIE 5632, light-emitting diode materials and devices, pp 324–331. doi:10.1117/12.581266

Burchell MJ, Graham G, Kearsley A (2006) Cosmic dust collection in aerogel. Annu Rev Earth Planet Sci 34:385–418. doi:10.1146/annurev.earth.34.031405.124939

Burschka J, Pellet N, Moon S-J, Humphry-Baker R, Gao P, Nazeeruddin MK, Grätzel M (2013) Sequential deposition as a route to high-performance perovskite-sensitized solar cells. Nature 499:316–319. doi:10.1038/nature12340

Carson S, Schmader J, Hanecek C, Gadley J. Biopolymers

Charrier T, Durand MJ, Affi M, Jouanneau S, Gezekel H, Thouand G (2010) Bacterial bioluminescent biosensor characterisation for on-line monitoring of heavy metals pollutions in waste water treatment plant effluents. In: Serra PA (ed) Biosensors. InTech, Rijek, pp 179–206. doi:10.5772/7210

Cho SH, Sung J, Hwang I, Kim RH, Choi YS, Jo SS, Lee TW, Park C (2012) High performance AC electroluminescence from colloidal quantum dot hybrids. Adv Mater 24:4540–4546. doi:10.1002/adma.201201524

Co Design (2015) This glowing disco ball could be your next clean water source. http://www.fastcodesign.com/3042136/this-glowing-disco-ball-could-be-your-next-clean-water-source. Accessed 13 Apr 2015

Coder KD (1999) Foxfire: bioluminescence in the forest. University of Georgia

DARPA (2012) Hiding in plain sight. http://www.darpa.mil/NewsEvents/Releases/2012/08/16a.aspx. Accessed 19 Nov 2013

Das R Electroactive polymers and devices 2013–2018: forecasts, technologies, players. IDTechEx. http://www.idtechex.com/research/reports/electroactive-polymers-and-devices-2013-2018-forecasts-technologies-players-000347.asp. Accessed 20 Nov 2013

Day JC Fireflies and glow-worms: an introduction. http://www.firefliesandglow-worms.co.uk/bioluminescent_reaction.html. Accessed 12 Mar 2013

Day JH (1968) Thermochromism of inorganic compounds. Chem Rev 68(6):649–657. doi:10.1021/cr60256a001

Dean J (2003) Recent biotechnology innovation is a bit fishy: a fluorescent pet. The Wall Street J. http://online.wsj.com/article/0,SB1052344997654016OO,00.html. Accessed 14 Mar 2013

Dezeen (2015) This glowing disco ball could be your next clean water source. http://www.fastcodesign.com/3042136/this-glowing-disco-ball-could-be-your-next-clean-water-source. Accessed 13 Apr 2015

Decker Y Homeostatic facade system. http://www.deckeryeadon.com/projects/HomeostaticFacadeSystem.html. Accessed 20 Nov 2013

Dodson B (2012) Flexible, high-strength polymer aerogels deliver 'super-insulation' properties. Gizmag. http://www.gizmag.com/polymer-aerogel-stronger-flexible-nasa/23955. Accessed 26 Nov 2013

Dornob—design ideas daily. Heat-activated paint for color-changing interior designs. http://dornob.com/heat-actived-paint-for-color-changing-interior-designs. Accessed 26 Nov 2013

Eglowli Cambridge. Bioluminescent street lamps. http://2010.igem.org/Team:Cambridge/Tools/Lighting. Accessed 13 Nov 2013

Electroluminescent Store, Protected Wallpaper. http://www.electroluminescent-shop.com/en/product/wallpaper. Accessed 18 Nov 2013

EMPA Materials Science and Technology. New aerogel-based plaster provides better insulation: space technology for old buildings. http://www.empa.ch/plugin/template/empa/3/124302/—/l=2/changeLang=true/lartid=124302/orga=/type=/theme=/bestellbar=/new_abt=/uacc=. Accessed 13 Nov 2013

Evangelho J (2014) How ViviTouch and their 'HD feel' technology is enhancing audio, gaming, and even adult toys. Forbes. http://www.forbes.com/sites/jasonevangelho/2014/01/20/how-vivitouch-and-their-hd-feel-technology-is-enhancing-audio-gaming-and-even-adult-toys/. Accessed 13 May 2015

Finkenstadt VL (2005) Natural polysaccharides as electroactive polymers. Appl Microbiol Biotechnol 67 (6):735–745

Finley B, Jacobs M (2010) Phase diagram of CO_2 (carbon dioxide). X axis is temperature in kelvin; Y axis is pressure in bar (2010). Wikimedia commons. https://commons.wikimedia.org/wiki/File:Carbon_dioxide_pressure-temperature_phase_diagram.svg. Accessed 25 July 25 2016

Franinovic K, Wille F, Gmachl M, Wingfield R, Kretzer M, Bisig D (2013) Actuated matter. In: Ng R, Patel S (eds) Performative materials in architecture and design. Intellect Ltd., Bristol, pp 212–213

Frank T, Widder EA (2002) Visual ecology and bioluminescence. NOAA: islands in the stream: exploring underwater oases. http://oceanexplorer.noaa.gov/explorations/02sab/background/biolum/biolum.html. Accessed 13 Mar 2013

Fraunhofer ISE (2014) New world record for solar cell efficiency at 46 % French-German cooperation confirms competitive advantage of European photovoltaic industry. Press Release 26/1. Bernin, France and Freiburg, Germany

'Glowing Plant.' http://glowingplant.com/

Glynn R, Karagkiozi Z, Themistocleous T (2013) Soft kinetic prototypes for adaptive architecture. In: Stacey M (ed) Prototyping architecture: the conference papers. Building Centre Trust, London, pp 422–442

Goodall C (2011) Bioplastics: an important component of global sustainability. Biome Bioplastics, Marchwood

GoodFellow. All the materials you need for scientific and industrial research and manufacturing. http://www.goodfellow.com/. Accessed 29 Nov 2013

Grätzel M (2001) Photoelectrochemical cells. Nature 414:338–344

Grätzel M (2003) Dye-sensitized solar cells. J Photochem Photobiol C Photochem Rev 4(2):145–153. doi:10.1016/S1389-5567(03)00026-1

Haddock SHD, Moline MA, Case JF (2010) Bioluminescence in the sea. Annu Rev Mar Sci 2:443–493. doi:10.1146/annurev-marine-120308-081028

Hallcrest LCR (1991) Handbook of thermochromic liquid crystal technology. Hallcrest, Glenville

Han Y-L (2013) Aerogel materials for aerospace. In: Zhang S, Zhao D (eds) Aerospace materials handbook. CRC Press, Boca Raton, pp 699–743

Hara K, Arakawa H (2003) Dye-sensitized solar cells. In: Luque A, Hegedus S (eds) Handbook of photovoltaic science and engineering. Wiley, Chichester. doi:10.1002/0470014008.ch15

Harnik E, Herbstein FH, Schmidt GMJ, Hirshfeld FL (1954) The structure of overcrowded aromatic compounds. Part I. A preliminary survey. J Chem Soc: 3288–3294. doi:10.1039/JR9540003288

Hart JA, Lenway SA, Murtha T (1999) A history of electroluminescent displays. University of Minnesota. http://www.indiana.edu/~hightech/fpd/papers/ELDs.html. Accessed 18 Nov 2013

Heikenfeld JC, Steckl AJ (2003) Inorganic EL displays at the crossroads. Inf Disp 19:20–25

Hennerdal L-O, Berggren M (2011) Picture-to-picture switching in full-color thermochromic paper displays. Appl Phys Lett 99(18):183303. doi:10.1063/1.3656972

Hickok Cole Architects (2013) The office building of the future. Architizer. http://architizer.com/projects/the-office-building-of-the-future/. Accessed 20 Nov 2013

Holcim Foundation for Sustainable Construction. Bioluminescent devices for zero-electricity lighting, Seville, Spain. Holcim awards for sustainable construction, 'Next Generation' 3rd prize Europe. http://epiteszforum.hu/files/HA11_EUR_NG3_WinnerPoster.pdf. Accessed 20 Nov 2013

Hough M, Dolbey R (1995) The plastics compendium, vol 1. Rapra Technology Ltd., Shawbury

Howes P, Laughlin Z (2012) Material matters: new materials in design. Black Dog Publishing, London

Hüsing N, Schubert U (2000) Aerogels. In: Ullmann's encyclopedia of industrial chemistry, 6th edn. Wiley VCH, Weinheim, pp 621–646

Ilievski FF, Mazzeo AD, Shepherd RF, Chen X, Whitesides GM (2011) Soft robotics for chemists. Angew Chem Int Ed 50:1890–1895. doi:10.1002/anie.201006464

Inventables. Temperature-sensitive glass. https://www.inventables.com. Accessed 26 Nov 2013

ITKE (Institute for Building Construction and Structural Design)—Prof. Dr.-Ing. Jan Knippers, University of Stuttgart, Germany; Faculty for Architecture and Urban Planning (2013) ArboSkin: durable and recyclable bioplastics facade mock-up. Press release ITKE University of Stuttgart

Jordi C (2011) Biomimetic airship driven by dielectric elastomer actuators. PhD dissertation, Swiss Federal Institute of Technology (ETH), Zurich

Jorgenson GV, Lee JC (1985) Thermochromic materials research for optical switching films. In: Proceedings of SPIE 0562, optical materials technology for energy efficiency and solar energy conversion IV, vol 2, pp 3–5. doi:10.1117/12.966280

Junghänel M (2007) Novel aqueous electrolyte films for hole conduction in dye sensitized solar cells and development of an electron transport model. PhD dissertation, Freie Universität Berlin

Kalyanasundaram K, Grätzel M (2009) Efficient dye-sensitized solar cells for direct conversion of sunlight to electricity. Mater Matter 4(4):88

Kamer Maker. 3D-print architecture. http://www.kamermaker.com/?page_id=2. Accessed 29 Nov 2013

Kinder L (2013) Glow in the dark ice cream made from jellyfish costs £140 a scoop. The telegraph. http://www.telegraph.co.uk/news/science/10442158/Glow-in-the-dark-ice-cream-made-from-jellyfish-costs-140-a-scoop.html. Accessed 13 May 2014

King CN (1996) Electroluminescent displays. J Soc Inf Disp 4(1):1–8. doi:10.1889/1.1984982

Kistler SS (1931) Coherent expanded aerogels and jellies. Nature 127:741. doi:10.1038/127741a0

Kobayashi M, Kikuchi D, Okamura H (2009) Imaging of ultraweak spontaneous photon emission from human body displaying diurnal rhythm. PLoS ONE 4(7): e6256. doi:10.1371/journal.pone.0006256

Koebel M, Rigacci A, Achard P (2012) Aerogel-based thermal superinsulation: an overview. J Sol-Gel Sci Technol 63(3):315–339. doi:10.1007/s10971-012-2792-9

Kofod G, Paajanen M, Bauer S (2006) Self-organized minimum-energy structures for dielectric elastomer actuators. Appl Phys A 85(2):141–143

Kooroshnia M (2013) Leuco dye-based thermochromic inks: recipes as a guide for designing textile surfaces. In: 13th AUTEX world textile conference

Kretzer M (2012) Shapeshift: soft dielectric electroactive polymers and architectural surfaces. In: Bhatia N, Sheppard L (eds) Bracket: goes soft. Actar, Barcelona, pp 80–83

Kretzer M (2013a) The materiability research. In: Gengnagel C, Kilian A, Nembrini J, Scheurer F (eds) Rethinking prototyping proceedings of the design modelling symposium Berlin 2013. Universität der Künste Berlin, Berlin, pp 205–215

Kretzer M (2013b) Information materials: smart material based architectural design. In: Stouffs R, Sariyildiz S (eds) eCAADe 2013: computation and performance conference proceedings. TU Delft, Delft, pp 615–623

Kretzer M (2014) Architecture in the era of accelerating change. In: Gerber D, Huang A, Sanchez J (eds) ACADIA 2014: design agency. Proceedings of the 34th annual conference of the association for computer aided design in architecture. University of Southern California, Los Angeles, pp 463–472

Kretzer M, Glynn R (2013) Material animation. In: Ng R, Patel S (eds) Performative materials in architecture and design. Intellect Ltd., Bristol, pp 34–35

Kretzer M, Rossi D (2012) ShapeShift. Leonardo 45 (5):480–481

Kretzer M, In J, Letkemann J, Jaskiewicz T (2013) Resinance: a (smart) material ecology. In: Beesley P, Khan O, Stacey M (eds) ACADIA 2013: adaptive architecture. Proceedings of the 33rd annual conference of the association for computer aided design in architecture. University of Waterloo, Cambridge, pp 137–146

Kulčar R, Friškovec M, Hauptman N, Vesel A, Gunde MK (2010) Colorimetric properties of reversible thermochromic printing inks. Dyes Pigm 86 (3):271–277. doi:10.1016/j.dyepig.2010.01.014

Lampert CM (2004) Chromogenic smart materials. Mater Today 7(3):28–35. doi:10.1016/S1369-7021(04)00123-3

Lee J (2008) Bioluminescence: the first 3000 years. J Siberian Fed Univ Biol 1(3):194–205

Leydecker S (2008) Nano materials: in architecture, interior architecture and design. Birkhäuser, Basel

Lochmatter P (2007) Development of a shell-like electroactive polymer (EAP) actuator. PhD dissertation, Swiss Federal Institute of Technology (ETH), Zurich

Loop.pH, Blumen Wallpaper. http://loop.ph/bin/view/Loop/BlumenWallpaper. Accessed 18 Nov 2013

Lorcks J (2006) Bioplastics: plants, raw materials, products. Fachagentur Nachwachsende Rohstoffe, Gülzow

Lötzsch D, Ruhmann R, Seeboth A (2013) Thermochromic biopolymer based on natural anthocyanidin dyes. Open J Polym Chem 3:43–47. doi:10.4236/ojpchem.2013.33009

Ma L, Zhan X (2013) Dye-sensitized solar cells (DSSCs). In: Hu W (ed) Organic optoelectronics, 1st edn. Wiley-VCH, Weinheim, pp 437–465. doi:10.1002/9783527653454.ch10

Mæhlum MA (2013) Solar cell efficiency world record set by sharp—44.4 %. Clean Tech. http://cleantechnica.com/2013/06/23/solar-cell-efficiency-world-record-set-by-sharp-44-4. Accessed 12 Nov 2013

Mauch RH (1996) Electroluminescence in thin films. Appl Surf Sci 92:589–597. doi:10.1016/0169-4332(95)00301-0

Mecklenburg M, Schuchardt A, Mishra YK, Kaps S, Adelung R, Lotnyk A, Kienle L, Schulte K (2012) Aerographite: ultra lightweight, flexible nanowall, carbon microtube material with outstanding mechanical performance. Adv Mater 24(26):3486–3490. doi:10.1002/adma.201200491

Meta Territory Studio. Soft Modular Pneumatic System (SMPS). http://www.daekwonpark.com/work%20folder/SMPS.html. Accessed 19 Nov 2013

Metamorphic Architecture. Fabcell. http://metamo.sfc.keio.ac.jp/project/fabcell/index.html. Accessed 26 Nov 2013

Miettunen K, Halme J, Lund P (2013) Metallic and plastic dye solar cells. WIREs Energy Environ 2 (1):104–120. doi:10.1002/wene.46

Mills JFD, Nyburg SC (1963) Thermochromism and related effects in Bixanthenylidenes and Bianthronylidenes. Part I. Crystal structure analyses. J Chem Soc: 308–321. doi:10.1039/JR9630000308

Miodownik M (2008) The time for thermochromics. Mater Today 11(11):6

Morin SA, Shepherd RF, Kwok SW, Stokes AA, Nemiroski A, Whitesides GM (2012) Camouflage and display for soft machines. Science 337:828–832. doi:10.1126/science.1222149

Mossé A, Kofod G, Thomsen MR (2011) Materializing a responsive interior: designing minimum energy structure based on dielectric elastomers. In: Conference proceedings, adaptive architecture, London

NASA Glenn Research Center—Technology. Mechanically strong, flexible polyimide aerogels. https://technology.grc.nasa.gov/featured-tech/aerogels.shtm. Accessed 15 Nov 2013

Nicola Burggraf. http://www.nicolaburggraf.com/. Accessed 13 Nov 2013

Ogrodnik W (2008) Use of color-changing pigments to detect wire and cable hazards. Wire J Int 41(4):150–155

Ono YA (1995) Electroluminescent displays. World Scientific, Singapore

Ow DW, Wood KV, Deluca M, De Wet JR, Helsinki DR, Howellt SH (1986) Transient and stable expression of the firefly luciferase gene in plant cells and transgenic plants. In: Science 14, vol 234(4778):856–859. doi:10.1126/science.234.4778.856

Parkin IP, Binions R, Piccirillo C, Blackman CS, Manning TD (2008) Thermochromic coatings for intelligent architectural glazing. J Nano Res 2:1–20. doi:10.4028/www.scientific.net/JNanoR.2.1

Perry B (2007) Bioluminescent fungi. Mycoweb. http://www.mykoweb.com/articles/BioluminescentFungi.html. Accessed 12 Mar 2013

Peters S (2011) Material revolution: sustainable and multi-purpose materials for design and architecture. Birkhäuser, Basel

Philips (2011) Bio-light. Philips: design futures. http://www.design.philips.com/philips/sites/philipsdesign/about/design/designportfolio/design_futures/bio_light.page. Accessed 13 Nov 2013

Pierre AC (2011) A history of aerogels. In: Aegerter MA, Leventis N, Koebel MM (eds) Aerogels handbook. Springer, London, pp 1–18

Pierre AC, Pajonk GM (2002) Chemistry of aerogels and their applications. Chem Rev 102(11):4243–4265

Pierre AC, Rigacci A (2011) SiO_2 aerogels. In: Aegerter MA, Leventis N, Koebel MM (eds) Aerogels handbook. Springer, London, pp 21–45

Poltrona Frau. Poltrona Frau is at the 'Coliseum of the XXI Century' http://www.poltronafrau.com/portal/page/portal/new/webpages/poltronafrau/press/r/home?p=id:251390&lang=en. Accessed 18 Nov 2013

RavenBrick. http://www.ravenbrick.com. Accessed 26 Nov 2013

Riether G (2011) AIA pavilion, a flexible system of cellular building blocks. V!RUS [online], No. 6. http://www.nomads.usp.br/virus/virus06/?sec=7&item=2&lang=en. Accessed 13 May 2015

Ritter A (2007) Smart materials in architecture, interior architecture and design. Birkhäuser, Basel

Roda A (2010) A history of bioluminescence and chemiluminescence from ancient times to the present. In: Road A (ed) Bioluminescence and chemiluminescence: past, present and future. Royal Society of Chemistry, Cambridge, pp 3–50. doi:10.1039/9781849732024-00001

Rossi D, Nagy Z, Schlüter A (2013) Soft pneumatics for sustainable architecture. In: International workshop on soft robotics and morphological computation, workshop digest

Rossiter J, Yap B, Conn A (2012) Biomimetic chromatophores for camouflage and soft active surfaces. Bioinspiration Biomimetics 7(3):5

Runyana N, Blazieb D (2010) Electroactive polymers actuators aid the quest for the 'Holy Braille' of tactile displays. Proc SPIE 7642:764207-1-12. doi:10.1117/12.847764

Sage I (2011) Thermochromic liquid crystals. Liq Cryst 38(11–12):1551–1561

Sarik J, Akinwande AI, Kymissis I (2011) A laboratory-based course in display technology. IEEE Trans Educ 54(2):314–319. doi:10.1109/TE.2010.2055242

Science Service Smithsonian (1961) Moving lighted image. http://scienceservice.si.edu/pages/046002.htm. Accessed 18 Nov 2013

Seeboth A, Ruhmann R, Mühling O (2010) Thermotropic and thermochromic polymer based materials for adaptive solar control. Materials 3(12):5143–5168. doi:10.3390/ma3125143

Seymour S (2009) Fashionable technology: the intersection of design, fashion, science and technology. Springer Vienna Architecture, Vienna

Sharp (2013) Sharp develops concentrator solar cell with world's highest conversion efficiency of 44.4 %. Sharp Press releases. http://sharp-world.com/corporate/news/130614.html. Accessed 10 Sept 2013

Shepherd RF, Ilievski F, Choi W, Morin SA, Stokes SA, Mazzeo AD, Chen X, Wang M, Whitesides GM (2011) Multigait soft robot. Proc Natl Acad Sci 108(51):20400–20403

Shepherd RF, Stokes AA, Freake J, Barber J, Snyder PW, Mazzeo AD, Cademartiri L, Morin SA, Whitesides GM (2013) Using explosions to power a soft robot. Angew Chem Int Ed 52:2892–2896. doi:10.1002/anie.201209540

Solaronix, Cell Kits. http://shop.solaronix.com/cell-kits.html. Accessed 12 Nov 2013

SRI International. Electroactive polymer 'artificial muscle'. http://www.sri.com/engage/products-solutions/epam. Accessed 19 Nov 2013

Stasiek JA, Kowalewski TA (2002) Thermochromic liquid crystals applied for heat transfer research. Opto-Electron Rev 10(1):1–10

Stevens ES (2002) The reemergence of bioplastics. In: Green plastics: an introduction to the new science of biodegradable plastics. Princeton University Press, Princeton, pp 104–134

Strozyk E. Thermochromic. http://www.elisastrozyk.de/seite/thermochromic/thermochrom.html. Accessed 26 Nov 2013

Sun H, Xu Z, Gao C (2013) Multifunctional, ultra-flyweight, synergistically assembled carbon aerogels. Adv Mater 25(18):2554–2560. doi:10.1002/adma.201204576

Tillotson TM, Hrubesh LW (1992) Transparent ultralow-density silica aerogels prepared by a two-step sol-gel process. J Non-Cryst Solids 145:44–50. doi:10.1016/S0022-3093(05)80427-2

UL Standards (2013) UL 94, the standard for safety of flammability of plastic materials for parts in devices and appliances testing. http://ulstandards.ul.com/standard/?id=94_6. Accessed 29 Nov 2013

Ulrich S (2009) Funktionide. Master thesis, HAWK Hildesheim

Van der Maas D, Meagher M, Abegg C, Huang J (2009) Thermochromic information surfaces: interactive visualization for architectural environments. In:

Computation: the new realm of architectural design: 27th eCAADe conference proceedings, pp 491–496
Wang H, Sun K, Tao F, Stacchiola DJ, Hu YH (2013) 3D honeycomb-like structured graphene and its high efficiency as a counter-electrode catalyst for dye-sensitized solar cells. Angew Chem Int Ed 52 (35):9210–9214. doi:10.1002/anie.201303497
Washam C (2010) Plastics go green. ChemMatters: 10–12
Wenger S (2010) Strategies to optimizing dye-sensitized solar cells: organic sensitizers, tandem device structures, and numerical device modeling. PhD dissertation, École Polytechnique Fédérale de Lausanne
White MA, LeBlanc M (1999) Thermochromism in commercial products. J Chem Educ 76(9):1201–1205. doi:10.1021/ed076p1201
Wilson TV (2007) How bioluminescence works. How stuff works. http://animals.howstuffworks.com/animal-facts/bioluminescence.htm. Accessed 25 July 2015
Wittgenstein L (1971) Tractatus logico-philosophicus, 2nd edn of new Translation. Routledge & Kegan Paul, London
Woods L, Flom L (Interviewer) (2004) The reality of experimental architecture: an interview with lebbeus woods. Carnegie online. July/Aug. http://www.carnegiemuseums.org/cmag/bk_issue/2004/julyaug/feature2.html. Accessed 16 Apr 2015

Towards a New Softness

5

5.1 Introduction and Book Summary

> The great extinction that occupies current human culture has swept away celebration of such transcendent, eternal qualities. If I stand on the floating piers of the Venice lagoon, amidst a withering biosphere, my posture shifts. The ground is yawning, viscous, inducing queasy vertigo. My legs unconsciously tense themselves, reptile brain-inflected posture tensed by the elastic meniscus underfoot. The shift of my own posture inverts a confident gaze, sending it outward. The enclosing function of architecture shifts from consuming the surroundings. A renewed task appears: constructing synthetic ground.
>
> Beesley (2010a, Hylozoic Ground, p. 21)

The core argument of this book is that *information materials* have the potential to fundamentally revolutionize the common notion of architecture in structural, functional, aesthetic, and especially ideological terms, indicating a new 'softness' in building. Since architecture however, represents a highly diverse and complex discipline that is as influential on its environment as it is influenced by it, predicting if and when such change will occur remains a matter of speculation. Hence, instead of making conjectures on the exact impact of *information materials* on architecture the main objective of this work is to promote a different kind of design and material thinking to overcome current limitations that slow down their progressive exploitation. The argument is narrated around four chapters, each building upon one another to conclude in what is referred to as an *information material literary,* a natural proficiency in anticipating the value of *information materials* for the future of architectural design.

Chapter 1 expands upon emerging technological, demographic, and socioanthropological trends and emphasizes the potential of adaptive spaces or buildings to address the quickly advancing nature of our environment. It highlights present tendencies but also shortcomings of applying *information materials* in architectural design, which for the time being are mainly focused on their intrinsic capability of sensing and responding to environmental influences. In line with the current sustainability discourse this is ought to reduce potential hardware complications, lower a building's maintenance requirements, and improve its overall efficiency in regards to energy consumption and waste

Chapter 2 sets up the book's theoretical framework. It is split into five parts elaborating on the steadily changing sense of materiality throughout the course of architectural history

M. Kretzer, *Information Materials*, DOI 10.1007/978-3-319-35150-6_5

until the present. The structure of the chapter and related subchapters follows a largely chronological order focusing on the ideological value of materials rather than their explicit technological development. This stresses their importance far beyond technical specifications alone. Nevertheless, material innovation and progress will obviously continue to have a strong impact on both the practice as well as the theory of architecture and therefore deserve continuous attention. Next to improving a building's performance and life cycle *information materials* might, in junction with high-strength materials, such as carbon fibers or other synthetic composites, like **Kevlar**, allow for the creation of much lighter and thinner constructions which could lead to a new formal language. The properties of a 30-cm brick wall could hence be replaced by multifunctional foils only a few millimeters thick, as KieranTimberlake have already successfully demonstrated with their SmartWrap project.[1] Yet, as much as such developments are related to the availability of a material's base substrates, the complexity of its production, and associated costs they are highly dependent on acceptance from the general public, which can be subject to trust, fashion, or cultural meaning.

Chapter 3 exemplifies a pedagogic approach for conveying a fundamental distinction of *information materials* from traditional materiality. Stressing the importance of hands-on experience and 'design through making' it suggests a process-oriented model, focusing on time and behavior as key elements for handling dynamic materials. Building upon the value of open exchange and access to related content the chapter demands the development of an *information material literacy*, which will encourage cross-disciplinary communication and collective progress. The method is illustrated through an educational framework, refined during numerous experimental student projects and workshops, as well as the *materiability research network*, an online platform that provides open access to *information materials* and related content. Since the inception of this idea in 2010 and the establishment of the network mid 2012, public interest in the area has grown substantially, especially among the academic community. Considering that within the next five years many of these students will be graduating and taking on positions within architectural design firms it can be assumed that they will bring along some of the knowledge and experience they have gathered on *information materials* during their studies. In junction with further technological progress an increase in adaptive architectural concepts can be expected.

Chapter 4 aims at further stressing the difference between traditional materials and *information materials* by emphasizing their deliberate synthetic creation and dynamic immediacy in response to external stimuli. The chapter presents eight material phenomena in terms of their historical development and usage, working principle, assembly, and (speculative) architectural application. The aim of the chapter is not to propose particular scenarios in respect to a certain materiality but rather to inspire and encourage the literate readers to develop their own visions and, through evolving skills for cross-disciplinary communication, become integral drivers for prospective material development. Since the selection of materials only covers a small portion within a constantly expanding field, the hope is that by providing detailed essays, including illustrated step-by-step fabrication procedures, the readers will not only learn to gain access to additional materials independently but also cultivate a more holistic and abstract sensibility in using *information materials* for future architectural design.

[1]More information on the project can be found in Sect. 2.7.4 Concepts for Information Materials Usage in Architecture.

5.2 Thinking Information Materials

As described earlier, the aim of this book is to promote an awareness of the difference in between *information materials* and traditional materials and provide a viable concept for their progressive application within architecture and related design disciplines. The following is an attempt to answer the questions that were posed in Sect. 1.6.1 *A Different Kind of Material Thinking*.

5.2.1 How Has Material Development and Usage Evolved in Architecture Over the Course of Time and How Has This Influenced the Practice and Education of the Architect?

In Fig. 5.1 the British materials engineer Michael Ashby analyzes various material developments and the change of their 'relative importance' over time according to four major categories, metals, polymers and elastomers, composites, and ceramics and glasses.

Ashby claims that the materials of prehistory were fairly primitive and that the first substantial technological leap came with the development of rudimentary thermochemistry around 5000 BC. This allowed "the extraction of, first, copper and bronze, then iron," hence initiating the long supremacy of metallic materials until their peak in the 1960s. Since then, with pressures for weight and cost savings, the rate of development of new metallic alloys has greatly slowed down and polymers, advanced composites, and high-performance ceramics have taken over (Ashby 2005, p. 5). Ashby's diagram very graphically depicts the history and advancement of materials and, although originating from an engineering background, can easily be mapped to changes in architecture and building. Especially the impact of steel, Portland cement, and sheet glass since the mid-nineteenth century and the shift towards an increasing demand for non-metallic, more sophisticated materials in the mid-twentieth century can be clearly observed.

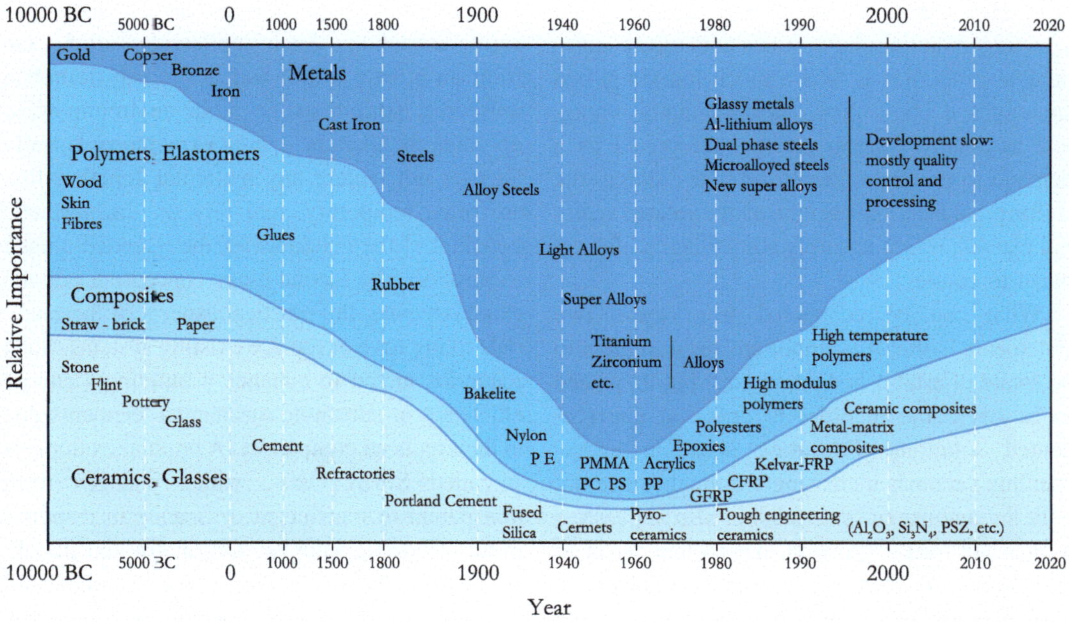

Fig. 5.1 Evolution of 'relative importance' of engineering materials over time (adapted from Ashby (2011), Copyright with permission from Elsevier)

Addressing the variety of today's materials, Michael Tomlan, a historic preservationist expert, observes that future material development will largely be affected by military needs, while their spread on the market will be determined by the economics of supply and demand. Emphasizing the importance of scientific experimentation to increase the range of possibilities he highlights the "need to understand the history and development of the building materials of the recent past" (Tomlan 1995, p. 43). Ashby, who calls the current time "the age of advanced materials," reasons that to make use of materials in an innovative way, designers need to resist the temptation to fall back into traditional applications simply because there is not enough guidance and information on novel alternatives (Ashby 2005, pp. 4, 8).

Both Ashby's diagram, as well as, Tomlan's suggestions in respect to approaching material-related design tasks are helpful in conveying a more general picture and potential technological tendency. Yet charts and statistics always have to be read with caution since no matter how scientifically grounded they might appear, they can only provide hints and indications but never guarantee universal validity. A purely scientific approach in categorizing and selecting materials based on standardized comparable properties is thus a much too simplified strategy and could prove fundamentally wrong, just as Ashby's graph can be interpreted as a demand to return to prehistoric times, due to the shifting balance and weight of the related material classes, which is surely something he did not mean to convey.

What can be concluded from these two statements is that it is important for designers to be aware of established paradigms, but they need to be overcome in order to arrive at unprecedented solutions. Educating such an understanding becomes increasingly difficult, since not only the amount of materials but also suggestive usages are vastly growing. This often leads to prematurely associating a certain material with a particular application, which reduces its potential spectrum. In the past, material culture constituted an elementary aspect of both architectural practice and education, which were closely linked and mutually dependent. Architects published their variegated knowledge in clearly structured treatises that were equally guidelines for building and primers for teaching. Today practice, education, and research have become largely separated areas and material-related issues are often excluded or isolated from design-centered approaches. Courses that communicate theoretical as well as practical skills while conveying a broad but comprehensive understanding of materiality can help students to develop a much more intuitive attitude towards materials so they can again become essential elements of design.

5.2.2 What Is the Current Meaning and Usage of Information Materials in Architecture and How Might It Change in the Future?

Currently there are two major tendencies when implementing dynamic materials in a spatial context. The first one focuses on emphasizing their transformative qualities due to their visual effects, turning them into spectacular and captivating attractions. Such applications often stem from an artistic origin and are mostly found in museums, exhibitions, or public environments to up-value the visitors' spatial experience. Usually they do not feature any particular functionality, are often temporary, and rely on the aesthetic potential of the materials for the intended artistic expression. The second usage concentrates on the exact opposite, the practical aspects of the materials to improve or replace existing systems. Such scenarios intend to enhance a building's energy efficiency or substitute mechanical elements due to their reduced complexity. A popular example is so-called Smart Glass, which changes from transparent to translucent or opaque in response to an electronic impulse. Originally and mainly applied in the car industry for automatically dimming side and rear mirrors, such window panes are increasingly finding their way into architectural space in order to eliminate the use of blinds for shading, privacy, or energy saving (Kolarevic and Klinger 2008, p. 9).

Oddly most new material technologies are combined with rather traditional materials, such as glass or even concrete, in order to make them architecturally sound. Next to Smart Glass examples include photovoltaic cells, **OLED** displays, **phase change materials** encased in glazing, or **self-healing concrete**. Obviously these decisions often stem from a manufacturer's understanding of architectural requirements towards materiality, aimed at making them more durable and lasting as well as relying on surfaces that are culturally and socially established. However such pragmatism also means discarding some of the materials' most supreme properties. Many of them can for example, instead of being sandwiched between two panes of glass, also made flexible using plastic films or foils. This would allow the construction of much thinner, lighter, and more complex, single or even doubly curved designs, applying techniques such as heat bending or thermoforming (Kretzer et al. 2013, p. 612).

Moreover the prevailing argumentation of not being able to apply *information materials* in architecture due to their limitations in scale and size might require reconsideration. As has been described in the *Intermediate Summary and Conclusion* of chapter two, architectural styles and with them the preferred usage of materials has constantly changed and alternated throughout history. Presuming a reemergence of sculptural and ornamental complexity, due to increasing advances in digital design and manufacturing technologies, the possible size of materials might not remain a determining factor for their architectural applicability. A focus on smaller elements in larger quantity could form spatial assemblies due to their collective amount and not their individual scale. Predicting the future usage of *information materials* in architecture, however, surely remains difficult. For the time being it might be centered on surface applications improving a building's efficiency in relation to energy saving or increased comfort and convenience. Yet with their growing occurrence in consumer products and related areas they could also permeate architecture in the form of smart home appliances, including building automation devices, entertainment systems, or domestic robots.

Their impact on architecture and many other aspects of the human environments is, however undeniable. Thus introducing architectural students to technological progress and material developments is extremely important and should mark a fundamental part of their education. With such knowledge in the minds of young professionals one can expect a rise in future design proposals of which technology becomes an integral part rather than just a means for form-generation, be it for the creation of new spaces or the adaptation of existing structures.

5.2.3 What Are Didactic Models that Focus on the Mediation of Dynamics, Processes, and Behaviors and How Can Explorative Material Research Become a Vital Aspect of Architectural Education?

In the opening of his *Ten Books on Architecture* Vitruvius emphasizes the requirement for an architect's diverse training and the mutual dependence of practice and theory. The purpose of Vitruvius' treatise, who admits, albeit having a rather broad knowledge, not being an expert in any of the arts, is thus exactly that, conveying an overview of everything that encompasses architecture equally of "importance not only to builders but also to all scholars" (Vitruvius 1914, pp. 5, 13). Physical making and the theoretical abstraction of it are thus inseparably linked, just as the immaterial is always connected to the material. When working with materials whose properties are changing dynamically over time it seems logical to draw from theories and educational systems, which emphasize thinking in terms of behaviors. Thomas Fisher refers to the

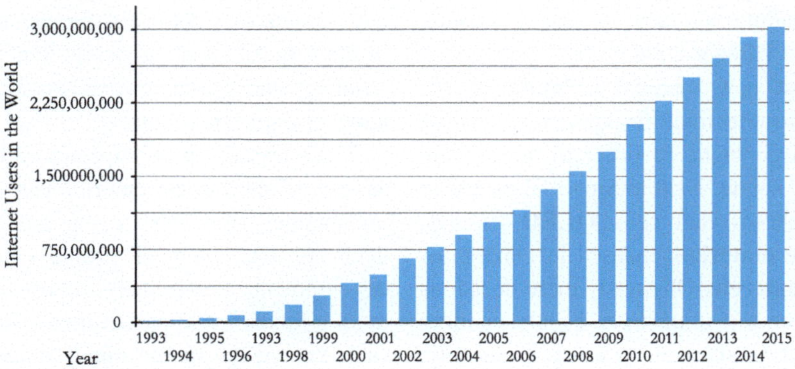

Fig. 5.2 Exponential growth of people having access to the Internet worldwide (adapted from Internet Live Stats, www.internetlivestats.com)

creative practice of Charles and Ray Eames when articulating a process-oriented model towards design where the path and related findings along the way are more important than the actual results. Moreover, he urges architects to see buildings not as isolated, everlasting elements but as spaces of continuity, changing and becoming with their occupants and associated events (Fisher 2000, pp. 46, 75).

The most prominent and omnipresent example of a dynamic, evolving, and self-improving didactic system is probably the Internet. As Fig. 5.2 shows, the amount of worldwide Internet users has grown exponentially from little more than fourteen million in 1993, making 0.002 % of the then world population, to over three billion today, representing almost 43 % of the global population.

The uniqueness of the Internet lies in its ability of connecting everything and everyone, independent of time or locality. Every thought, every idea, every event one shares, every place one goes, and every person who likes, comments, or even looks at it, are all linked, increasing ceaselessly. To thrive in such a quickly expanding environment and to extract valuable from irrelevant information cannot be taught but has to be learned by continuous usage. The more one engages in the process the more meaningful are the results one gets. Value is thus created on a metalevel, free from established paradigms, but simply through the amount of cross-links and interconnections. Quality becomes a direct result of quantity. Mapping such findings onto architectural education and particularly the mediation of emerging materials means to provide unrestricted and open access to related information, not teaching students 'what' to do but providing them with a stable and open framework for learning 'how' to do, advancing independence and self-exploration. While such a system certainly requires careful curatorship and guidance in the beginning it will, once being established and having reached a critical mass, continue to evolve autonomously, both in terms of quantity and quality.

5.2.4 How Can a Distinct Pedagogic Approach Be Evaluated and How Can Research Results Be Disseminated to a Large, Global Audience?

The success of the aforementioned approach, in the particular example of the *materiability research network*, can simply be measured by the number and frequency of user engagement, like for instance the amount of daily page views or the count of new versus returning visitors. More important than 'how' such results can be distributed technically, which in this case is more or less self-explanatory, is however the question 'if' they can be spread and what legal frameworks are necessary and appropriate. Unfortunately most of our world, including scientific disciplines, is still competition and profit driven. Research results are hence often held back until intellectual property rights, such as patents, have been claimed. Obviously issues such as authorship and credits are highly important and need to be addressed appropriately, yet they can also substantially slow down or even prevent collective work, especially when it aims at joining areas whose understandings and values might differ.

One possibility to bridge the gap in between the radical openness of the Internet and the binding rules of our legal and social systems are so-called Creative Commons licenses. The idea of developing an infrastructure, which allows public sharing while retaining individual property, emerged at the turn of the century as an alternative to established copyright systems strongly tied to the fundamentals of the Open Source Initiative. With free redistribution and access to information being one of the keystones of the Open Source philosophy it rises through the collaborative efforts of a multitude of users that build upon each other's work and continue developing it within specific communities (Raymond 2001, p. 26). Creative Commons empowers the creators of such and any other work to clearly, and legally protected, define usage, distribution, and editing rights within a standardized and globally applicable framework. As such it is flexibly situated in between full copyright and the public domain, as graphically visualized in Fig. 5.3, allowing a work for example to be shared and used for educational purposes, however excluding it from profit-making contexts (Aliprandi 2011, pp. 13–14).

Fig. 5.3 Creative Commons is flexibly situated between the "all rights reserved" model (classical copyright) and the "no rights reserved" idea (public domain) (adapted from Aliprandi 2011)

5.3 Liquid Architecture

> Liquid architecture is an architecture that breathes, pulses, leaps as one form and lands as another… it is an architecture without doors and hallways, where the next room is always where I need it to be and what I need it to be.
>
> Novak (1991, Liquid Architectures in Cyberspace, p. 250)

The dream of a transformable, responsive, and soft architecture as an alternative and in contrast to the classical notion of building, which is still largely based on the Vitruvian sense of stability, rigidity, and long-term usefulness, kept and keeps resurfacing. Often such concepts seem to be tightly linked to new technological developments, leading to euphoric and highly speculative visionary designs, yet they also tend to arise from a general, more subliminal dissatisfaction with the status quo.

In 1966 published David Greene, a founding member of Archigram, his Living Pod, a conceptual proposal for a kinetically adaptable, nomadic dwelling that could function autonomously or in unison with larger systems. Greene believed that due to permanent technological progress and a growing importance of personal mobility "the need for a house (in the form of a permanent static container) as part of man's psychological makeup will disappear." Hence, besides the Living Pod's futuristic form and advanced technical infrastructure, which was imagined to include self-leveling compression legs, wash capsules with automatic body cleaning equipment, or a food dispenser with self-cooking capability, its true radicality lay in suggesting a novel concept of the 'house' as the center of a person's private environment. Through this Greene not only questioned the common understanding of what architecture should be and what it stands for, such as security and permanence, but also promoted a viable and serious alternative, turning it into a form of enhanced clothing for a new type of urban nomads within an increasingly fluctuating world.

> With apologies to the master, the house is an appliance for carrying with you, the city is a machine for plugging into (Cook 1999, p. 52).

Whereas Greene and his contemporaries, while demanding a reconsideration of the essentials of architecture, were still largely bound by the laws of the physical world, came with the rise of computer-generated 3-D graphics and the Internet a completely new dimension, freed from time and space: Virtual Reality (Mitchell 1996, p. 8). Strongly motivated by these unprecedented possibilities developed Marcos Novak, professor at the University of California, Santa Barbara and director of transLAB, in the early 1990s his concept of 'liquid architectures.' Novak describes liquid architecture as continuously changing, evolving, and becoming, whose true and final identity can only be experienced over the full course of its existence. Situated at the threshold of **cyberspace** and reality, liquid architecture is based on motion and movement, mediating and uniting the virtual and the physical world. Novak's work invites the user to immerse in constructed, artificial environments that react and mutate in response to their own programmed behavior and a person's individual involvement. By this liquid architecture becomes more than an alternative or substitution to existing structures but rather an extension, adding the possibilities of information technology to develop a notion of space that fuses information, art, and architecture. Novak's vision of space is four-dimensional, endlessly being dematerialized and reassembled, with time constituting an essential factor of its creation and appearance.

> Liquid architecture is more than kinetic architecture, robotic architecture, and architecture of fixed parts and variable links. Liquid architecture is an architecture that breathes, pulses, leaps as one form and lands as another (Novak 1991, pp. 250–251).

Notwithstanding the vast availability of such ideas that emphasize the value of space as an emotional extension of the human body most of today's architecture that claims kinetic transformability is largely focused on technical functionality and based on rather simple mechanical principles. Prominent examples like Almann Sattler Wappner's Herz Jesu Kirche (Fig. 5.4) in Munich (2000), Wes Jones' Sci-Arc Conference Room (2003), DRMM's Sliding House (2009), or Aedas' Al Bahr Towers (2013) in Dubai are simply scaled-up replica of well-established machinery, celebrated more for their visual effect than fundamental architectural novelty.

Yet they still encounter tremendous technical problems since structure, surface, and actuators remain separate entities and since the necessary equipment often has vast spatial and acoustical impacts. This results in spaces and systems that have highly mechanistic characteristics and which, due to their structural complexity, are intense in operation and service (Kretzer 2011).

Information materials have advantages over such methods in both technological and conceptional terms. By unifying structure, skin, actuator, and in cases even sensor within one system they avoid mechanical complexity and thus save costs, energy, and maintenance requirement. More interesting, however, is the subliminal impact they have on people. Since their dynamic behavior is occurring at scales impossible for the human eye to perceive and

Fig. 5.4 The huge front doors of the Herz Jesu Kirche in Munich are opened on special holidays and merge the interior of the building with the public space (Kretzer 2004)

since their performance is often nonlinear and thus hard to anticipate, they impose something magical and utterly organic. If one was to neglect for a moment the cultural importance of existing architecture and likewise ignore sustainability in terms of reduction as the prime demand of any future building but instead assume the industrial production and availability of *information materials*, such materiality would allow the conception of a fundamentally different type of architecture. Situated at the verge of the digital and the physical, *information materials* could enable a combination of both Greene's and Novak's visions, becoming active expansions of the human body that provide existential support while materially mediating between the virtual and the real.

Obviously such a vision, as has been stated earlier, is still rather far-fetched and despite constant progress in various scientific areas towards 'living' environments, including research into synthetic biology or artificial intelligence, might require more than only a new type of materiality to succeed.

5.4 Towards a New Softness

> You're not a beautiful and unique snowflake. You're the same decaying organic matter as everything else. We're all part of the same compost heap. We're all singing, all dancing crap of the world.
>
> Palahniuk (2010, Fight Club, p. 134)

Returning to the philosophy of Lucretius, as described in Chap. 2, to whom all matter is in constant flux, it is interesting to observe that the crucial aspect of time as an additional dimension of architectural materiality is hardly ever addressed, neither in theoretical nor in more scientific positions. Even though nature is often cited as a source or inspiration, architecture is generally represented as some immutable, static entity, unaffected and independent of decay or death, ought to remain as imagined for as long as possible. Yet if one looks at the role of materials beyond their physical attributes but as crucial parts of our human environment, as extensions of our physical realm and bodies, they cannot simply be considered as infinite entities. Just as all life is a process of constant adaptation and change, materiality should also be considered in a more dynamic context, with time representing a fundamental element to be integrated into architectural design.

In his book *Camouflage* architect and theorist Neil Leach builds upon the notion of adaptation and discusses the desire and primal condition of humans to either assimilate and blend into their surroundings "as a vital mode of survival in an ever-changing world," and as such a defense mechanism, or to use similar strategies in order to distinguish themselves from the average. Elaborating upon this phenomenon and the concomitant process of "becoming one with the world, and of becoming distinct from that world" he emphasizes the importance of architectural place-identity and the social role design plays in enabling the successful operation of human camouflage. Describing the core feature of the human species as continually evolving new strategies to engage with each other and the environment, Leach demands new architectural models that allow rethinking our relationship with place and space. Camouflage is portrayed by Leach as an aesthetic tool that facilitates the formation of such a relationship, demanding a focus on the ongoing and temporal connection between architecture and the human body and as such a correlation between matter (the inanimate) and life (Leach 2006, pp. 3–5, 7, 244–245).

The immersive responsive installations of the Canadian architect and artist Philip Beesley build upon a similar motivation, trying to evoke a sensual, emphatic link between the visitor and the artwork (Beesley and Elsworthy 2006, p. 169). In contrast to most architectural work, is Beesley's approach centered on the philosophical notion of Hylozoism, a belief that all matter is animate and inseparably connected to life. His artworks or "textiles have [...] taken the form of interlinking matrices of mechanical components and arrays of sensors and actuators that respond to occupants moving within the environment" (Beesley 2010a, b, p. 19). The expansive

Fig. 5.5 Kim Taehyung in Philip Beesley's Epiphyte Chamber, exhibited at the Museum of Modern and Contemporary Art, Seoul, Korea (Beesley Architect Inc. 2013)

installations, which in cases bridge across several rooms, are made from millions of minute laser-cut plastic elements that snap-fit into artificial spatial sculptures reminiscent of plastic forests, creating a synthetic nature (Fig. 5.5).

The intricate structures are equipped with countless embedded electronics, tracking and actively responding to visitors, as well as a myriad of glands and bladders, filled with vaporous chemicals that invigorate the space and its guests. Beesley's works are an oxymoron, through their acrylic materiality they could not be more artificial and yet they form highly naturalized environments that appear so coherent that the visitors, the only truly living entities, feel like intruders accidentally having entered a parallel world. The emotional quality of this approach in both seducing the observers (who might very well be the observed) while actively isolating them becomes a vivid mirror of the paradox of our time, that is defined by a deep desire to return to more naturalized environments on the one hand while synthesizing our daily life on the other (Kretzer 2014, p. 19). Beesley's work is unsettling and disturbing yet reassuring and elusively calm. It urges us to reassess our relationship to the ground and, in a time marked by fluctuation and ephemerality, makes us wonder about the meaning of stability and solidity.

The prime demand in Beesley's projects, particularly the Hylozoic series, to embody empathy, marks a deliberate antipole to the present and dominant conception of architecture (Beesley 2007, p. 20). Due to this and their excessively fragile constitution, they reveal an immanent softness, a biosynthetic malleability, embracing yet revealing, sheltering yet conjoining. Being immersed within one of his installations, which he personally describes as 'geotextiles,' no distinction can be felt in between structure and skin, form and function, or even interior and exterior. Through this amalgamation of qualities he succeeds in overcoming one of the core problems of adaptive systems, which are usually added on top of existing structures that albeit intended to offer increased flexibility instead consume valuable space for their technical infrastructure.

Just as the transdisciplinary practice of Philip Beesley, suggest *information materials* the arrival of architecture at a 'new softness' due to a number of very unique possibilities. As mentioned before, many of these materials are thin, flexible, pliable, or even literally based on organic matter, making them appear and feel much softer than traditional materiality, such as steel, concrete, or glass. These physical properties enable the design and creation of fluid forms, which, promoted further by their specific behaviors, break out of persistent paradigms and instead focus on the inverse, ephemerality, change, and variation. Due to their contrast to mechanical systems, *information materials* hence allow the natural mediation between soft environments, such as the interior and the exterior space of a building, or in between the sensual boundaries of humans.

To visualize such emotive qualities two projects that both focused on color changing thermochromics are compared for their subliminal impact, *Resinance*, realized by the CAAD MAS class at ETH Zürich in 2013 and *Dynamics in Extreme Environments*, a workshop at the

Institute for Advanced Architecture of Catalonia in 2015.

During the latter the group *Thermoteam* explored the temperature-induced effect of thermochromics for the creation of a dynamic double layer facade system, which should be activated by external sun radiation or an integrated heat ventilation system. The active skin was intended to display varying colors and through this change the building's appearance and internal atmosphere based on the temperature difference between the inside and the outside of the structure. Even though such a system would certainly be very aesthetic and might decrease the necessity for mechanical shading, it does highlight a rather pragmatic approach, replacing an established technology within an existing framework. Obviously the students only worked on this proposal for a few days and in that respect achieved outstanding results, yet employing *information materials* to substitute a mechanical system, represents a mechanistic way of thinking and hence not only reduces the materials far below what they are capable of but also employs them as mechanical devices.

Resinance on the other hand neglected any kind of obviously useful functionality. The installation consisted of forty hollow elements, made from thermochromic plastic. Each container was filled with water, which when heated lead to a slow and gradual color change of the objects' surfaces. Moreover, all elements were networked together so that a change in one unit's status also affected all its neighbors, leading to a successive evolution of the complete installation. Through this very subtle but constant mutation, paired with occasional shivers when physically touched, the project developed a certain organic animism, suggesting qualities of living environments. So even though its immediate effect was comparably little, *Resinance* was able to impact visitors on a very sensual level, turning the applied materiality into much more than just a color changing substrate.

Engaging in implications rather than the actual effects might hence be crucial in order to develop truly revolutionary spatial experiences and situations. An interesting setting to explore would be the private home, which Jonathan Hill describes as "the one place that is considered to be truly personal," and as such providing both physical and emotional stability (Hill 2006, p. 8). Due to the continuously shifting nature of our personal and social environment Hill argues that the home must remain stable and comforting, a place that defines the threshold between interior (personal) and exterior (public) world. The home thus marks a critical archetype for wondering about the larger consequences a dynamic materiality would have in terms of spatial physicality and the resulting social and cultural repercussions.

In 1962 the English novelist James G. Ballard published a dystopian short story entitled *The Thousand Dreams of Stellavista*, describing a 'psychotropic house' that, in a machine-like manner, is able to respond to and learn from its occupants. Through its occupation and continuous exchange with its tenants the building begins to adapt its material structure and consequently evolve its own personality over time. Consisting of a substance that Ballard refers to as plastex, a mixture of plaster and latex, the house is able to change and adjust its shape as desired. An excerpt from the novel, when the new tenants Mr. and Mrs. Talbot are first shown the apartment by their agent Stamers, describes the particular characteristics of the material as follows:

> "Lively responses, aren't they, Mr. Talbot?" He put his hand on the wall behind us. The plastex swam and whirled like boiling toothpaste, then extruded itself into a small ledge. Stamers sat down on the lip, which quickly expanded to match the contours of his body, providing back and arm rests (Ballard 1971, p. 187).

Despite the tragic ending of the story during which the house, enabled through its senso-cells that allow it to read every physical and mental change of its occupants, attempts to kill them in an act of despair and traumatic memories, the notion of a fully responsive, materially adapting building raises a number of interesting considerations which often tend to be dismissed for the sake of technophiliac progress.

- What impact would dynamic materials have on the way we perceive space?
- How could we orient ourselves in an environment, which is constantly changing and how fast can such change happen without causing confusion?
- How can active materials be applied in a way that grounds rather than leading to further estrangement from the natural world?
- How will the cultural and social notions of stability and consistence change?
- How much control are we willing to surrender and how do we establish trust in such a system?

Certainly the above list is far from comprehensive and only highlights a brief selection of related issues. Returning to the claim of Lucretius and considering the current age of technological advancement as being at the verge of creating a second nature, facilitated through progress in nanotechnology, genetic engineering, and synthetic biology, they are, however, aspects worth considering when speculating about future architectural scenarios. Yet, while adaptivity might empower a space to change based on a person's individuality it could also result in the exact opposite, utterly depersonalized environments, for the sake of remaining adaptable, similar to the tendency in digital fabrication where progress in machining flexibility requires increasingly standardized materiality. Employing *information materials* in a farsighted and responsible manner will thus not only involve a careful evaluation of all involved parameters but especially design visions that dare to question and look beyond immediate horizons towards the needs and desires of tomorrow's world.

I like to think (and
  the sooner the better!)
  of a cybernetic meadow
  where mammals and computers
  live together in mutually
  programming harmony
  like pure water
  touching clear sky.

I like to think
  (right now, please!)
  of a cybernetic forest
  filled with pines and electronics
  where deer stroll peacefully
  past computers
  as if they were flowers
  with spinning blossoms.

I like to think
  (it has to be!)
  of a cybernetic ecology
  where we are free of our labors
  and joined back to nature,
  returned to our mammal
  brothers and sisters,
  and all watched over
  by machines of loving grace.

Brautigan (1967, All Watched Over By Machines Of Loving Grace, p. 126)

References

Aliprandi S (2011) Creative commons: a user guide. Ledizioni, Milano

Ashby MF (2005) Materials selection in mechanical design. Butterworth-Heinemann, Amsterdam

Ashby MF (2011) Introduction. In: Materials selection in mechanical design, 4th edn. Butterworth-Heinemann, Amsterdam, pp 1–13

Ballard JG (1971) The thousand dreams of Stellavista. In: Vermilion sands. Vintage Random House, London, pp 185–208

Beesley P, Elsworthy W (2006) Implant matrix. In: Beesley P, Hirosue S, Ruxton J, Tränkle M, Turner C (eds) Responsive architectures: subtle technologies 2006. Riverside Architectural Press, Cambridge, pp 168–171

Beesley P (2007) Hylozoic soil: geotextile installations 1995–2007. Riverside Architectural Press, Cambridge

Beesley P (2010a) Hylozoic ground: liminal responsive architecture. Riverside Architectural Press, Cambridge

Beesley P (2010b) Kinetic architectures & geotextile installations. Riverside Architectural Press, Toronto

Brautigan R (1989) Richard Brautigan's Trout fishing in America, the pill versus the Springhill mine disaster,

and in watermelon sugar. Houghton Mifflin/Seymour Lawrence, Boston
Cook P (1999) Archigram. Princeton Architectural Press, New York
Fisher T (2000) In the scheme of things: alternative thinking on the practice of architecture. University of Minnesota Press, Minneapolis
Hill J (2006) Immaterial architecture. Routledge, London
Kolarevic B, Klinger KR (2008) Manufacturing material effects: rethinking design and making in architecture. Routledge, New York
Kretzer M (2011) Towards a new softness. In: Stacey M (ed) Adaptive architecture international conference proceedings, DVD conference papers. The Building Centre, London
Kretzer M, Minuto A, Nijholt A (2013) smart material interfaces: another step to a material future. In: proceedings ICMI'13: 15th ACM—international conference on multimodal interaction, ACM, Sheridan, Sydney, 9–13 December 2013, pp 611–612
Kretzer M (2014) Spaces of adaptivity: thoughts on the relationship of life and architecture. In: Kretzer M, Hovestadt L (eds) ALIVE: advancements in adaptive architecture. Birkhäuser, Basel, pp 16–23
Leach N (2006) Camouflage. MIT Press, Cambridge
Mitchell W (1996) City of bits: space, place, and the infobahn. MIT Press, Cambridge
Novak M (1991) Liquid architectures in cyberspace. In: Benedikt M (ed) Cyberspace: first steps. MIT Press, London, pp 225–254
Palahniuk C (2010) Fight club. Vintage, London
Raymond ES (2001) The cathedral & the bazaar, 2nd edn. O'Reilly Media, Sebastopol
Tomlan M (1995) Building modern America. In: Jester TC (ed) Twentieth-century building materials: history and conservation. McGraw-Hill, New York, pp 34–43
Vitruvius P (1914) The ten books on architecture (trans: Morgan MH). Harvard University Press, Cambridge

Appendix A

A.1 Credits

The projects, experiments, and tutorials presented in this volume emerged from intense collaborative and cross-disciplinary exchange and are built upon the vast experience of countless professionals in the various fields. The relentless support and dedication of formerly unknown individuals was not only extremely encouraging, and literally essential for achieving the final results, but also stands exemplary for a general openness and interest in engaging in each other's domain to collectively progress. In particular, thanks go to all the highly skilled students and workshop participants who tirelessly advanced the respective tasks, arriving at novel and inspiring proposals for the future of architectural design and education. Additional credit is due to Prof. Ludger Hovestadt at the Chair for Computer Aided Architectural Design, ETH Zurich, for providing infrastructure, knowledge, funding, and especially mental and personal support throughout the course of this endeavor. All credits are arranged according to their chronological appearance in the text.

A.1.1 Material Credits

Aerogels

The instructions for a granular aerogel were developed with great support from Matthias Koebel, Shanyu Zhao, and Jeong Eunho at the Building and Nanomaterials—Group, Laboratory for Building Science and Technology, Swiss Federal Laboratories for Materials Science and Technology (EMPA) and documented together with Achilleas Xydis. Special thanks goes to Shanyu Zhao who was extremely supportive cross-reading the chapter and providing essential insights and comments.

Bioluminescence

The tutorial for growing bioluminescent algae was developed with help from Marin Sawa (http://www.marins.co.uk/) and was made possible through extended conversations with Maike Lorenz of the Culture Collection of Algae at Goettingen University.

Bioplastics

The procedure on making bioplastics is based on instructions found in E. S. Stevens' book *Green Plastics: An Introduction to the New Science of Biodegradable Plastics* and was developed with the great help of Swiss designer Beat Karrer (http://www.beatkarrer.com/), who has worked with biopolymers for several years. It was documented during the 2011/12 MAS CAAD module *Molecular Utopia* together with Miro Roman and Evangelos Pantazis.

Dye-Sensitized Solar Cells

The instructions for thin and flexible dye-sensitized solar cells were developed with

M. Kretzer, *Information Materials*, DOI 10.1007/978-3-319-35150-6

great support from Paul Liska of Michael Grätzel's Laboratory of Photonics and Interfaces (LPI) at the École Polytechnique Fédérale de Lausanne (EPFL), which has improved and developed dye-sensitized solar cells since the early 1990s, as well as John Meschter of G24 innovations, a UK-based company that specializes in the fabrication of advanced, flexible thin film dye-sensitized solar cells. They were documented during the 2011/12 MAS module *Molecular Utopia* with help from Miro Roman and Evangelos Pantazis. Paul Liska deserves special credit for reading the chapter and verifying its content.

Electroluminescent Displays

The fabrication process for electroluminescent displays was adapted with help by John Sarik from the Columbia University course Modern Display Science and Technology, a class developed by researchers at the Columbia Laboratory for Unconventional Electronics (CLUE). CLUE uses thin film materials and low temperature fabrication techniques to build sensors, displays, and other electronic devices on unconventional substrates. This novel integration enables the creation of large, flexible, nonplanar electronic systems. The procedure was documented during the 2011/12 MAS CAAD module *Molecular Utopia* with help from Miro Roman and Evangelos Pantazis. Particular credit is due to Emil Enz from the Swiss electroluminescent manufacturer Lumitec AG (http://www.lumitec.ch/) for proofreading the chapter.

Electroactive Polymers

The fabrication procedure on electroactive polymers was made possible through the extensive research into electroactive polymers technology of the group around Gabor Kovacs, Mechanical Systems Engineering at the Swiss Federal Laboratories for Materials Science and Technology (EMPA). Special thanks goes to Christa Jordi of the same group for her endless support in explaining the procedure and helping in solving details and technical problems. The instructions were documented during the 2011/12 MAS module *Molecular Utopia* with the help of Miro Roman and Evangelos Pantazis. Gabor Kovacs deserves particular credit for providing his lab space for intense experimentation and ensuring the accuracy of the provided information.

Soft Robotics

The tutorial on soft robotics is based on research that was originally developed at the Whitesides Research Group, Harvard University Department of Chemistry and Chemical Biology. It was made possible and documented through an exploratory workshop initiated by Roman Kirschner and the *Liquid Things* research project and was led by Ben Haworth, who spent a great amount of time developing soft actuators during his architectural thesis at the MSc Adaptive Architecture and Computation, Bartlett UCL.

Thermochromics

The procedure explaining the creation of a hollow thermochromic plastic container by mixing leuco dye pigments with polyester resin was supported by Sara Robertson, Lecturer in Craft Innovation/Smart Materials at the University of Dundee, United Kingdom who has vast experience in working with thermochromic materials. The process was developed together with Hironori Yoshida and documented during the 2012/13 MAS CAAD module *Material Ecologies* by Achilleas Xydis and Demetris Shammas.

A.1.2 Project Credits

Bioplastic Formfinding Workshop

The *Bioplastic Formfinding Workshop* took place at the Architecture Biennale PLAN12 in Cologne. It was organized by Mauricio Rodriguez (MAS ETH) and co-tutored by Hans Sachs (http://responsivedesign.de/). It was supported by the Materiability Research Network, Superartitecture, Dingfabrik e.V. (FabLab Cologne) and was held at the exhibition space and studio of Cologne based design and architecture office Superartitecture.

Phototropia

The project was realized within four weeks by the 2011/12 MAS class at the Chair for CAAD (Katia Ageeva, Diana Alvarez, Orestis Argyropoulos, Stella Azariadi, Tianyi Chen, Yun-Ying Chiu, Ivana Damjanovic, García Pepo Martínez, Melina Mezari, Bojana Miskeljin, Evangelos Pantazis, Stanislava Predojevic, Stylianos Psaltis, Meda Radovanovic, Daniel Rohlek, Miro Roman, Castro Mauricio Rodríguez, Teemu Seppänen, Grete Soosalu). It involved the creation of electroactive polymers, electroluminescent displays, dye-sensitized solar cells, and bioplastics. The electroactive polymer team was tutored by Florian Wille (ZHdK IAD), the electroluminescent group was supervised by Luke Franzke (ZHdK IAD), the bioplastics were explained by Beat Karrer and the principles of dye-sensitized solar cells presented by Paul Liska (EPFL LPI) and John Meschter (G24 innovations). In addition Andrei Pruteanu and Agostino di Figlia (TU Delft ES group) joined the project for one week in order to program the responsive behavior of the installation together with a small group of students who dealt with solving the electronic circuitry and power conversion from one device to the other. Special thanks goes to Karmen Franinovic.

Material Animation

The project was realized within four weeks by the 2010/11 MAS class at the Chair for CAAD (Agata Muszynska, Aleksandar Lalovic, Hideaki Takenaga, Jesper Thøger Christensen, Jorge Orozco, Magda Osinska, Mihye An, Nikola Marincic). The course was co-tutored by Ruairi Glynn (Bartlett, UCL) and supported through Emil Enz (Lumitec AG), Romano Kirschbaumer and Jorge Ellert (Ulano Corp.), John Sarik (CLUE), Florian Wille, Luke Franzke, Karmen Franinovic, and Max Rheiner (Zurich University of the Arts, ZHdK), and Klaus Wassermann from the Chair for CAAD.

Actuated Matter Workshop

The *Actuated Matter Workshop* took place at the Zurich University of the Arts from July 25 to 29, 2011. It was supported through a generous fund from ZHdK and organized together with by Karmen Franinovic and Florian Wile (ZHdK, IAD), Mathias Gmachl and Rachel Wingfield (http://loop.ph/), and Daniel Bisig (ZHdK, DMU). The workshop involved twenty international participants (Katrin Bächli, Urban Bieri, Szilveszter Buzasi, Allison Dryer, Luke Franzke, Laura Kaehr, Moritz Kemper, Roman Kirschner, Jorge Orozco, Barbara Peikert, Margrit Rieben, Maria Smigielska, Andrés Villa Torres, Silvan Zurbruegg), who developed a speculative model for membrane structures that exhibited properties of sensitivity, resilience, and decay. It was further supported by Romano Kirschbaumer and Jorge Ellert (Ulano Corp.), and Christa Jordi (EMPA)

Luminous Textiles

The one-week workshop in fall 2015 was organized together with Delia Dumitrescu, at the Swedish School of Textiles in Boras. Participants included Riikka Townsend, Evelin Kägo, Jyoti Kapur, Elias Högberg, Sofie Di Bartolomeo, Lisa Andersson, Vidmina Stasiulyte, and Marjan Kooroshnia. Special thanks goes to Delia Dumitrescu.

ShapeShift

ShapeShift was the final MAS CAAD thesis project of Edyta Augustynowicz, Sofia Georgakopoulou, Dino Rossi, and Stefanie Sixt and was realized in September 2010. The project investigated the use of dielectric elastomers to create dynamic spatial applications. The project was greatly supported by Dr. Gabor Kovacs, Christa Jordi, Christian Dueragar, Angelo Scioscia, and Sebastian Valet (EMPA—Swiss Laboratories for Materials Science and Technology), and Leonhard Fünfschilling (IKEA Stiftung Schweiz). In particular, thanks go to Gabor Kovaks for allowing to use his lab environment and associated tools and materials and especially to Christa Jordi for her endless support and dedication in progressing and successfully completing the project.

Animated Textiles

The one-week workshop in fall 2012 was held together with Ivana Damjanovic, at the Swedish School of Textiles in Boras, invited by Delia Dumitrescu. The participants included Astrid Mody, Delia Dumitrescu, Felicia Davis, Una Baldvinsdottir, Joanne Kowalski, Inese Parkova, Emelie Johansson, Riikka Saarela, Christina Maschke, Stella Katsarou, Nilla Berko, and Justien De Bus. Special thanks goes to Delia Dumitrescu, Florian Wille, and Karmen Franinovic.

Soft Motion and Liquids

Soft Motion and Liquids was a three-day workshop held as part of the *Liquid Things* research project at the University of Applied Arts (Angewandte), Vienna from Nov. 30th to Dec. 3rd 2012. The workshop was initiated and organized by Roman Kirschner. It was co-organized by Ben Haworth and was based on his thesis work into soft robotic system at the MSc Adaptive Architecture and Computation, Bartlett UCL, with additional support from Ruairi Glynn. The workshop built upon predefined shapes that Ben Haworth had produced together with Zoi Karagiozi. Participants included Ben Haworth (Bartlett School of Architecture), Ruairi Glynn (Bartlett), Manuel Kretzer (ETH, CAAD), Marcel Finke and Roman Kirschner (Liquid Things), Anita Peretti, Solmaz Farhang, Isidora Krstic, Alin Cirstea, Kostas Politis.

Resinance

The project was realized within four weeks by the 2012/13 MAS class at the Chair for CAAD (Baldwin Mark, In Jessica, Janjusevic Tihomir, Jiang Nan, Letkemann Joel, Miranda Turu Nicolás, Prieler Irene, Schildberger David, Shammas Demetris, Smigielska Maria, Tanigaito Aki, Xexaki Evi, Xydis Achilleas, Yuko Ishizu). The course was co-tutored by Benjamin Dillenburger and Hironori Yoshida (ETH CAAD), Weixin Huang and Lei Yu (Tsinghua University, China), Tomasz Jaskiewicz and Mariana Popescu (Hyperbody, TU Delft, the Netherlands), and Andrei Pruteanu and Stefan Dulman (Embedded Software Group, TU Delft, the Netherlands).

Resinance 2.0

Resinance 2.0 was the successor of *Resinance*, realized six month after the first project. The project emerged from a student application to showcase the work at the *2013 ACADIA conference* at the school of architecture, University of Waterloo, Cambridge, Canada. The design and material research was performed by Achilleas Xydis and the electronics and interaction design was developed by Joel Letkemann. Further collaborators included Demetris Shammas, Evi Xexaki, Maria Smigielska, Mariana Popescu, Nan Jiang, and Yuko Ishizu. ACADIA support came from Farzin Asad, Zak Fish, and Connor O'Grady. The installation was set up and installed on site by Achilleas Xydis.

Dynamics in Extreme Environments

Dynamics in Extreme Environments was a four-day workshop held at the Institute of Advanced Architecture of Catalonia as a collaboration between IaaC, ETH CAAD and Chalmers Technical University, exploring a selection of materials for the creation of architectural proposals in extreme environments. The course was co-supervised by Areti Markopoulou (IaaC - Institute for Advanced Architecture of Catalonia) and Stig Anton Nielsen (Chalmers Technical University, Dept. of Architecture) with assistance from Alexandre Dubor and Carlos Bausa (IaaC - Institute for Advanced Architecture of Catalonia). The students were split into five teams. *Thermoteam* involved Alina Kvirkveliya, Ji Won Jun, Josep Alcover, Jose Algeciras, Matteo Silverio, and Yoko Gotoh. *Survival Bubble* consisted of Maragkoudaki Anna, Tsung Hsun Ku, Devika Chandrababu, Ninada Kashyap, and Fathimath Sujna Shakir. *Thermoluminescence* was comprised of Nina Jotanoivc, Panagiota Sarantinoudi, Farah Alayeli, Apostolos Apostolinas, Lubna Alayeli, and Ceren Temel. *Maneuvering Reticulation* was the project of Christoph Zechmeister, Daphne Arnellou, Despoina Pippa, Luisa Roth, Nohelia Gonzalez, and Shreyas More. *Delaying Desertification* was developed by Anusha Arunkumar, Eirini Aikaterini Papakonstantinou, Kunal Chadha, Mansi Jasmin Radadia, Ryal Sequeira, Silasalin Krisanarungkhun, and Yessica Gabriela Mendez Sierra.

A.2 Material Suppliers

The following selection of materials and ingredients refers to the respective tutorials as described in the chapter *A Selection of Emerging Information Materials, their Properties, Fabrication, and Application in Speculative Spatial Installations*. The listed materials represent the then core sources, which were chosen due to local constraints or other reasons, yet due to the quickly advancing nature of *information materials* long-term validity cannot be ensured. Material suppliers, descriptions, volumes, quantities, prices, availability, and even names might have changed in the meantime. The following list means to provide an overview and give hints on where to find the right sources but does not guarantee completion. In case descriptions have changed please do *not* contact the author nor the publisher but feel invited to post your findings on the platform www.materiability.com and simplify the search for others.

Aerogels

- Hydrochloric acid (HCl), Sigma-Aldrich, Hydrochloric acid solution 31088 FLUKA, 305.50 CHF/1 l (12. 2013), http://www.sigmaaldrich.com/catalog/product/fluka/31088?lang=de&region=CH
- Hexamethyldisiloxane (HMDSO), Sigma-Aldrich, Hexamethyldisiloxane ≥ 98 %, 25 CHF/100 ml (12. 2013), http://www.sigmaaldrich.com/catalog/product/aldrich/205389?lang=de&region=CH
- Precursor solution (PEDS-P75E20), Produits Chimiques Auxiliaires et de Synthese SA (PCAS), P75E20 (pre-hydrolysed TEOS with 20 % w/w of SiO_2), http://www.pcas.com/advanced_materials_precursors.html
- N-heptane, Sigma-Aldrich, n-Heptane 592579 SIGMA-ALDRICH, 62.60 CHF/1 l (12. 2013), http://www.sigmaaldrich.com/catalog/product/sial/592579?lang=de&region=CH
- Ethanol, Sigma-Aldrich, Ethanol 02860 FLUKA, 49.80 CHF/25 ml (12. 2013), http://www.sigmaaldrich.com/catalog/product/fluka/02860?lang=de&region=CH
- Amberlyst 15 (wet) hydrogen form ion exchange resin, Sigma-Aldrich, Amberlyst® 15 hydrogen form 216399 ALDRICH, 47.70 CHF/25 g (12. 2013), http://www.sigmaaldrich.com/catalog/product/aldrich/216399?lang=de&region=CH
- Water glass (sodium silicate) solution, Sigma-Aldrich, Sodium silicate solution 338443 SIGMA-ALDRICH, 30.40 CHF/25 ml (12. 2013), http://www.sigmaaldrich.com/catalog/product/sial/338443?lang=de&region=CH
- 5.5 mol/l ammonium hydroxide, Sigma-Aldrich, Ammonium hydroxide solution 318612 FLUKA, 31.90 CHF/500 ml (12. 2013), http://www.sigmaaldrich.com/catalog/product/fluka/318612?lang=de&region=CH.

Bioluminescence

- Pyrocystis Lunula, Culture Collection of Algae at Goettingen University (SAG), Pyrocystis Lunula 2014, 16 EUR/10 g (09. 2013), http://sagdb.uni-goettingen.de/detailedList.php?str_number=2014
- Micro Algae Grow fertilizer, Florida Aqua Farms Inc., Micro Algae Grow, 15 USD/2.6 lbs (09. 2013), http://florida-aqua-farms.com/shop/micro-algae-grow.

Bioplastics

- Gelatin powder, Landi Nidwalden AG, Speisegelatine, 34.50 CHF/1 kg (09. 2013), http://www.landinidwalden.ch/html/speisegelatine_17001.aspx
- Glycerol, pharmacy or drug store, 19.90 CHF/0.5 l, (01. 2015).

Dye-sensitized Solar Cells

- Transparent ITO Coated Film F2071018D1, Gwent Group Advanced Material Systems, 2.50 EUR/A4 sheet (11. 2012), http://www.gwent.org/gem_data_sheets/polymer_systems_products/electroluminescent_display_materials/ito_film_f2071018d1.pdf Dye Sensitized Solar Cell Kit, Mansolar: Article 1000 Kit, 49.50 EUR (03. 2015), http://www.mansolar.nl/products/article-1000-kit-1.html, Institute for Chemical Education: Nanocrystalline Solar Cell Kit, 45 USD (03. 2015), http://ice.chem.wisc.edu/Catalog/SciKits.html#Anchor-Nanocrystalline-41703, Sol Ideas: Nanocrystalline Dye-Sensitized Solar Cell Kit, http://www.solideas.com/solrcell/ICE_98_001_NanocrystalllineSolarCellKit.html.

Electroluminescent Displays

- Silhouette cutting machine, Silhouette SD Digital Craft Cutter, 260.50 CHF (09. 2013), http://www.overstock.com/Crafts-Sewing/Silhouette-SD-Digital-Craft-Cutter-with-10-Gift-Card/4400810/product.html
- Ulano knife cut films, Ulano, Sta-Sharp S3S Knifecut Film, http://www.ulano.com/knifecut/stasharp.htm, or UlanoCut Green, http://www.ulano.com/knifecut/ulanocut.htm
- ITO (indium tin oxide) coated foil, Gwent Group Advanced Material Systems, Transparent ITO Coated Film F2071018D1, 2.50 EUR/A4 sheet (11. 2012), http://www.gwent.org/gem_data_sheets/polymer_systems_products/electroluminescent_display_materials/ito_film_f2071018d1.pdf
- Phosphor paste, Gwent Group Advanced Material Systems, High Brightness Phosphor Ink C2070209P5, 927.30 EUR/1 kg (11. 2012), http://www.gwent.org/gem_data_sheets/polymer_systems_products/electroluminescent_display_materials/green_phosphor_c2070209p5.pdf
- Dielectric paste, Gwent Group Advanced Material Systems, White Dielectric Paste D2070209P6, 277.37 EUR/1 kg (11. 2012), http://www.gwent.org/gem_data_sheets/polymer_systems_products/electroluminescent_display_materials/dielectric_paste_white_d2070209p6.pdf
- Silver paste, Gwent Group Advanced Material Systems, Silver Ink C2090210P12, 1161.96 EUR/1 kg (11. 2012), http://www.gwent.org/gem_data_sheets/polymer_systems_products/electroluminescent_display_materials/silver_ink_c2090210p12.pdf
- EL inverter, Enz Electronic, EL-Inverter, E0406010 (200 cm^2): 4 CHF, 20406015 (500 cm^2): 60 CHF, E0410031 (2500 cm^2): 250 CHF, (11. 2012), http://www.enz.ch/pages/de/el-inverter.php
- Electro Luminescent Foils, Lumitec AG, Multi A5: 75 CHF, Multi A3: 170 CHF, (11. 2012), http://www.lumitec.ch/d/folien/multi/#A5
- EL Experimenter Kit, Gwent Electronic Materials Ltd, http://www.gwent.org/gem_electroluminescent_kit.html.

Electroactive Polymers

- Elastomeric membrane foil (VHB 4910 or VHB 4905), IBZ AG, VHB 4910 F33m, 10442.SR8135,
  1291.40 CHF/115 mm × 66 m (09. 2013), http://www.ibzag.ch/de/home.html
- Graphite powder carbon black, Akzo Nobel, Ketjenblack EC-330JMA, http://www.akzonobel.com/polymer/our_products/ketjenblack_ec_330jma/index.aspx

- 5–5000 V power converter (G50R or G60R), Condatas AG, VDC Converter, G50R (145 CHF) G60R (150 CHF), (11. 2012), http://www.condatas.com/dynasite.cfm?dsmid=104751
- Polycarbonate, Vink AG, Lexan PC Folie 0.25mm, 14.20 CHF/m^2, (11. 2012) http://www.vink.ch/
- Liquid Silicone, Suter Kunststoffe, NEUKASIL RTV 23, http://www.swiss-composite.ch/
- High Voltage Resistors, 4 × 10 MΩ
- Foam rubber, Maagtechnic, Platte PRENA Zellkautschuk 10113565, 45.70 CHF (11. 2012), http://www.maagtechnic.ch/en_CH.html.

Soft Robotics

- Soft silicone rubber (Eco Flex 00-30), Smooth-On, Ecoflex Supersoft Silicone, 183.72 USD/16 lbs (09. 2013), http://www.smooth-on.com/Silicone-Rubber-an/c2_1115_1130/index.html
- Vacuum chamber, Vacuum Degassing, 3 Litre Vacuum Degassing Chamber Kit - Model DP-3-Kit, http://www.vacuum-degassing.com/degassing-kits.html
- Hard silicone rubber (Addition Cure 33), Tiranti, Addition Cure 33 Silicone Rubber 1 kg kit, 25.85 GBP/1 kg (09. 2013), http://www.tiranti.co.uk/edgeimpactshop/product.php?Product=2957
- Phosphorous powder, Glo Nation, Super Phos Green Glow Pigment, 89.95 USD/16 oz (09. 2013), http://glonation.com/glow-powder.php.

Thermochromics

- Polyester resin, Suter Swiss Composite, 18536.0001, 16.50 CHF/kg (03. 2015), http://www.swiss-composite.ch/
- Thermochromic pigments, Chromazone, http://www.chromazone.co.uk/Powder.htm, Amazon, 29.99 USD/0.3 oz, http://www.amazon.com/Thermochromic-Pigment-Changes-Color-Temperature/dp/B009E9NCW6
- Catalyst hardener, Suter Swiss Composite, SCS M20 MEKP Härter (5 %) 145.1304, 122.75 CHF/5 kg (03. 2015), http://www.swiss-composite.ch/
- Cobalt Activator, Suter Swiss Composite, 10100.0100, 14.80 CHF/100 ml (03. 2015), http://www.swiss-composite.ch/.

Glossary

Adenosine Triphosphate (ATP) transports chemical energy within all living cells for metabolism. The energy in ATP is obtained from the breakdown of foods

Arduino is an open-source computer hardware and software project that produces microcontroller kits for building physical computing devices and interactive objects. For programming the boards, the Arduino platform offers its own integrated development environment (IDE) based on another open-source project: Processing

Basidiomycota are filamentous fungi including the groups: mushrooms, puffballs, stinkhorns, bracket fungi, other polypores, jelly fungi, boletes, chanterelles, earth stars, smuts, bunts, rusts, mirror yeasts, and the human pathogenic yeast Cryptococcus

Biocrude is also known as biooil or pyrolysis oil. It is a synthetic fuel made from dried biomass (plant material). It is compatible with existing refinery technology and can be converted into advanced biofuels or used for the production of plastic materials. Biocrude plastics cannot be used directly but through compounding are mixed with additives or aggregates to meet specific requirements

Biohacking or wetware hacking involves a wide range of practices and movements from "grinders" who design and install do-it-yourself body-enhancements such as magnetic implants in their fingers to biologists who perform at-home gene sequencing. Biohacking emerged in a rising trend of noninstitutional science and technology progress

Bragg Diffraction occurs when electromagnetic radiation encounters an obstacle and must scatter in various directions. Bragg's law allows for the angles at which the beam of electromagnetic radiation is scattered when the beam strikes a solid crystalline surface to be known

CCD (charge-coupled device) is a device for capturing the movement of electrical charge, used mainly in digital imaging. Although CCDs are not the only technology for light detection, CCD image sensors are widely used in professional, medical, and scientific applications where high-quality image data is required

Chemical Vapor Deposition (CVD) is a process to create high-quality, high-performance, solid materials, often used in the semiconductor industry to produce thin films

Cherenkov Counters When a charged particle passes through a transparent medium with a velocity larger than the phase velocity of light in the medium, it emits prompt photons, called Cherenkov radiation. Cherenkov

M. Kretzer, *Information Materials*, DOI 10.1007/978-3-319-35150-6 counters are particle detectors that make use of this radiation

Cybernetics is an approach for exploring regulatory systems, their structures, constraints, and possibilities, focusing on the design or discovery and application of principles of regulation and communication. Cybernetics treats not things but ways of behaving. It is applicable when the analyzed system generates change in its environment and that change is reflected in the system in a manner that leads to a system change

Cyberspace as a new term was first introduced by science-fiction author William Gibson in his 1982 short story 'Burning Chrome' and again in his novel 'Neuromancer,' published in 1984. It aims at describing a virtual environment in which communication through and in between computer networks takes place, such as the Internet

Doctor Blade is used in printing and coating processes. The doctor blade removes the excess ink from the smooth non-engraved portions of an image carrier

Dopant can also be called a doping agent. It is a trace impurity element that is introduced into a substance (in very low concentrations) to change the electrical or optical properties of the substance

Escherichia coli usually abbreviated *E. coli*, are a Gram-negative, rod-shaped bacterium that occur in the lower intestine of warm-blooded organisms. *E. coli* are a large and diverse group of bacteria and most strains of *E. coli* are harmless

Endocrine Disruptors are chemicals that can interfere with a mammal's endocrine system and lead to adverse developmental, reproductive, neurological, and immune effects. A large number of substances, both natural and synthetic, may cause endocrine disruption, including many everyday products such as plastic bottles, metal food cans, detergents, flame retardants, food, toys, cosmetics, and pesticides

Erythrosine is a cherry-pink synthetic compound, mainly used for food coloring. It is an organic compound containing iodine and sodium and produced from coal tar

Fab Labs are basically small to medium sized community spaces that provide access to various digitally controlled machines such as laser-cutters, 3D printers, CNC milling machines, and often basic electronic prototyping facilities

G-code is the common name for the most widely used numerical control (NC) programming language. It is used mainly in computer-aided manufacturing for controlling automated machine tools

Gutta Percha is a tough plastic substance from the latex of several Malaysian trees of the sapodilla family. It is similar to rubber but contains more resin and is often used as insulation and in dentistry for temporary fillings

Gram-Negative bacteria are a class of bacteria that do not retain the crystal violet stain used in the Gram staining method of bacterial differentiation. Gram-negative bacteria cause infections including pneumonia, bloodstream infections, wound, or surgical site infections, and meningitis in healthcare settings. Gram-negative bacteria are resistant to multiple drugs and are increasingly becoming resistant to most available antibiotics

Invertebrates The animal kingdom is informally divided into two groups, the vertebrates and invertebrates. Invertebrates have no backbone, unlike animals such as reptiles, amphibians, fish, birds, and mammals

Kevlar is the trademark name for a para-aramid synthetic fiber, developed by DuPont. The high-strength material has many applications, from bicycle tires and racing sails to body armor, due to its high tensile strength-to-weight ratio, which is five times more than steel

Kickstarter is a global crowd-funding platform whose aim is to bring realize and fund creative projects, such as films, music, stage shows, comics, journalism, video games, and food-related ideas

Microelectromechanical Systems (MEMS) is the technology of very small devices made up of components between 1 to 100 micrometers in size, usually consisting of a central microprocessor and several components that interact with the environment

Monocoque is a structural approach through which loads are carried through an object's external skin, similar to an egg shell. The technique is also referred to as structural skin. Monocoques are especially in the aircraft and automobile racing industry of increased interest in order to reduce weight and increase safety

N-region, P-region A p–n junction is a boundary or interface between two types of semiconductor material, p-type and n-type, inside a single crystal of semiconductor. It is created by doping, for example by ion implantation, diffusion of dopants, or by epitaxy (growing a layer of crystal doped with one type of dopant on top of a layer of crystal doped with another type of dopant)

Organic Light-Emitting Diode (OLED) is a light-emitting diode (LED) in which the electroluminescent layer is a film of organic compound which emits light in response to an electric current

Phase-Change Material (PCM) is a substance which by recursively melting and solidifying at programmed temperatures, can store and release substantial amounts of energy. In architecture they are often used as latent heat storage devices in floors or facades

Physical Vapor Deposition (PVD) describes a variety of vacuum deposition methods used to deposit thin films by the condensation of a vaporized form of the desired film material onto various workpiece surfaces (e.g., onto semiconductor wafers)

Redox Mediator is a hyperbranched polymer attached to an electrode and assisting in transferring electrons between the electrode and a redox enzyme

Self-healing Concrete is a new research area that aims at developing cement-based materials which are capable of repairing themselves after being damaged. Attempts include embedding shape-memory materials, adding sunlight-activated protective coatings, or even introducing bacteria that multiply and fill cracks when activated

Supercritical Point In thermodynamics, a critical point (or critical state) is the end point of a phase equilibrium curve. The most prominent example is the liquid–vapor critical point, the end point of the pressure–temperature curve that designates conditions under which a liquid and its vapor can coexist

Team 10 was a group of architects who assembled during the CIAM IX meeting in 1953 and who challenged the normative modernist approach to architecture and urbanism. Their theoretical agenda, spread mainly through teaching and publications, posed a profound influence on the progress of architectural thinking during the second part of the twentieth century

Wattle and Daub is a composite building material used for making walls, in which a woven lattice of wooden strips called wattle is daubed with a sticky material usually made of some combination of wet soil, clay, sand, animal dung, and straw

XBee are radio modules that can be attached to an Arduino to achieve wireless communication or the exchange of data among adjacent devices

The manufacturer's authorised representative in the EU is Springer Nature Customer Service Centre GmbH, Europaplatz 3, 69115 Heidelberg, Germany. If you have any concerns regarding our products, please contact ProductSafety@springernature.com

Printed and bound by CPI Group (UK) Ltd, Croydon, CR0 4YY
06/10/2026
02228904-0001